AIR UNIVERSITY

AIR FORCE RESEARCH INSTITUTE

Ideas in Arms

The Relationship of Kinetic and Ideological Means in America's Global War on Terror

Thomas D. Torkelson
Lieutenant Colonel, USAF

Drew Paper No. 2

Air University Press
Maxwell Air Force Base, Alabama 36112-5962

[redacted]

Muir S. Fairchild Research Information Center Cataloging Data

Torkelson, Thomas D.
Ideas in arms : the relationship of kinetic and ideological means in America's global war on terror / Thomas D. Torkelson.
p. ; cm. – (Drew paper, 1941-3785 ; no. 2)
Includes bibliographical references.
ISBN 978-1-58566-178-7.
1. Strategy. 2. War on Terrorism, 2001- 3. United States – Operations other than war. 4. Information warfare. 5. Military art and science – United States. I. Title. II. Series.
355.02—dc22

This Drew Paper and others in the series are available electronically at the Air University Research Web site http://research.maxwell.af.mil and the AU Press Web site http://aupress.maxwell.af.mil.

The Drew Papers

The Drew Papers are occasional publications sponsored by the Air Force Research Institute (AFRI), Maxwell AFB, Alabama. AFRI publishes this series of papers to commemorate the distinguished career of Col Dennis "Denny" Drew, USAF, retired. In 30 years at Air University, Colonel Drew served on the Air Command and Staff College faculty, directed the Airpower Research Institute, and served as dean, associate dean, and professor of military strategy at the School of Advanced Air and Space Studies, Maxwell AFB. Colonel Drew is one of the Air Force's most extensively published authors and an international speaker in high demand. He has lectured to over 100,000 students at Air University as well as to foreign military audiences. In 1985 he received the Muir S. Fairchild Award for outstanding contributions to Air University. In 2003 Queen Beatrix of the Netherlands made him a Knight in the Order of Orange-Nassau for his contributions to education in the Royal Netherlands Air Force.

The Drew Papers are dedicated to promoting the understanding of air and space power theory and application. These studies are published by the Air University Press and broadly distributed throughout the US Air Force, the Department of Defense, and other governmental organizations, as well as to leading scholars, selected institutions of higher learning, public-policy institutes, and the media.

All military members and civilian employees assigned to Air University are invited to contribute unclassified manuscripts that deal with air and/or space power history, theory, doctrine, or strategy, or with joint or combined service matters bearing on the application of air and/or space power.

Authors should submit three copies of a double-spaced, typed manuscript and an electronic version of the manuscript on removable media along with a brief (200-word maximum) abstract. The electronic file should be compatible with Microsoft Windows and Microsoft Word—Air University Press uses Word as its standard word-processing program.

Please send inquiries or comments to
Director
Air Force Research Institute
155 N. Twining St.
Maxwell AFB, AL 36112-6026
Tel: (334) 953-9587
DSN 493-9587
Fax: (334) 953-6739
DSN 493-6739
E-mail: research.support@maxwell.af.mil

Contents

Illustrations

Tables

Foreword

Lt Col Thomas D. Torkelson claims that the inability of the United States to achieve its stated political objectives in its global war on terror (GWOT) reflects its flawed kinetic-centric military strategy. This study erects a framework of effectiveness utilizing Clausewitzian principles to judge military strategy. By considering the expressed political objectives of the GWOT, the centers of gravity (COG) that military strategy should target within this struggle, and the GWOT's placement along a Clausewitzian continuum of violence, this paper evaluates US military efforts in Operation Enduring Freedom (OEF) through the perspective of two opposing models. First, the paper presents a kinetic model founded in America's historical implementation of its military arm that emphasizes physical effects. Second, it constructs a nonkinetic model comprised of the primarily psychological components of the nation's informational capabilities. Examining actions in OEF through these two lenses reveals the institutional tension the US military experienced as it sought to leverage the relevant COG.

Stemming from the preeminence of ideology among the COGs of GWOT, the paper finds that the US military and civilian leadership came to recognize the detrimental effects of its continued emphasis upon kinetic means in OEF. As a thesis and antithesis conflict in a Hegelian, dialectic fashion, the US military arrived at a synthesis for its operations in Afghanistan. Acknowledging the need to elevate nonkinetic campaigns as the primary effort in OEF, US military operations began to stress actions directed at the COG of ideology. The author concludes that although the ultimate outcome of the GWOT remains to be determined, positive institutional experiences such as the provincial reconstruction team (PRT) concept need to be preserved, while significant national public diplomacy reform must occur.

Ideas in Arms: The Relationship of Kinetic and Ideological Means in America's Global War on Terror was originally written as a master's thesis for the Air University's School of Advanced

Air and Space Studies (SAASS) at Maxwell AFB, Alabama, in June 2007. The Airpower Research Institute is pleased to publish this SAASS research as a Drew Paper and thereby make it available to a wider audience within the US Air Force and beyond.

DANIEL R. MORTENSEN
Chief of Research
Airpower Research Institute

About the Author

Lt Col Thomas D. Torkelson is a 1992 graduate of the US Air Force Academy in Colorado Springs, Colorado, where he earned a bachelor of science degree in political science. He earned his pilot wings at Reese AFB, Texas, in September 1993, graduating in February 1994 from KC-135R initial qualification training at Castle AFB, California, as a distinguished graduate. His first operational assignment to the 909th Air Refueling Squadron, Kadena AB, Japan, was highlighted by his upgrade to aircraft commander as a first lieutenant. In September 1996, he received a humanitarian reassignment to the 12th Airlift Flight, Langley AFB, Virginia, to fly the C-21A. While at Langley, he upgraded to instructor and evaluator pilot and was a personal instructor for the commander, Air Combat Command. Returning to Kadena in April 2000, he upgraded to instructor and evaluator pilot in the KC-135R while at the same time completing a master of arts degree in economics from the University of Oklahoma. He served as chief, Standardization/Evaluation, KC-135 Branch, at the 18th Operations Group before being reassigned to the 615th Air Mobility Operations Squadron at Travis AFB, California, in April 2003, where he served as deputy chief, Strategy Division. He also attended the KC-135 Weapons Instructor Course as a part of Class 04A, where he earned top flyer, top academics, and distinguished graduate distinctions. He then attended the Air Force Institute of Technology in May 2005 for Intermediate Developmental Education where he earned a master's degree in logistics management as well as distinguished graduate honors. A 2007 graduate of the School of Advanced Air and Space Studies at Maxwell AFB, Alabama, Colonel Torkelson now serves in the Strategy, Plans, and Policy Division of United States Northern Command, Peterson AFB, Colorado.

Acknowledgments

To God, with Him one can do all things. To Debbie, a paragon of patience and the epitome of selflessness, you are the world's greatest wife and mother. I love you. To Jack, thanks for making third grade look easy. To Adam, thank you for all of the wrestling breaks. To Jonah, thank you for all your hugs and kisses. To Catherine Carole, thank you for gracing our family this year with such beautiful femininity.

To my classmates at SAASS, particularly Tim "Ace" Kirk and George "Drool" Holland, thank you for tolerating my often inane comments in the pursuit of knowledge.

Special gratitude is reserved for Dr. Stephen Chiabotti for inspiring this topic and for Dr. Phillip LaSala for providing graciously his insight and direction.

Col William M. Darley, USA, also offered valuable contributions to this study. I appreciate sincerely his attention and willingness to participate.

Finally, to Lt Col Dave "Tilta" Werling, thank you for many instances of wry humor, intelligent debate, and mutual respect. XVI is not the same without you, but we are all better off for having known you. See you on the other side. . . .

Chapter 1

The Timeliness of Timelessness

War should be waged not for the goal of victory, necessary though that usually is, but rather for the securing of an advantageous peace.

—Colin S. Gray

The dawn of an information age powered by exponential increases in computing capability enraptures America's national security community. Emerging during a time when the United States sought an identifiable adversary, the information revolution provided some degree of direction for national defense planning. Whether under the moniker of a revolution in military affairs (RMA) or defense transformation, the US military has embraced the notion of information as an engine of change. Captivated by the allure of near-perfect situational awareness, near-instant information dissemination, and ever-shortening decision-making cycles, the US military presently pursues technological advances to enhance its already formidable prowess to conduct conventional kinetic operations rather than to elevate its capabilities within wars of ideas. Operation Desert Storm seemed to presage the wisdom of this force structure strategy. However, in the actions that followed the horrific and spectacular attacks upon the World Trade Center and the Pentagon in 2001, America's traditional kinetic emphasis in its military operations has proven insufficient to achieve desired political objectives. Drawn by the information revolution's timely appeal, the American military seems to have forgotten the timeless power of information and knowledge in warfare.

From Sun Tzu in the centuries before Christ, to Carl von Clausewitz of the Napoleonic Age, and on to Colin Gray and others today, most serious studies of warfare acknowledge the central importance of information and the force of ideas. Even the Melian Dialogue of the Peloponnesian War in the fifth century B.C. reveals the instrumental role of information and discussion in warfare despite the massacre that followed.[1] Although the United States has recognized the value of information in warfare, its history reflects a propensity to pursue strategies of technologically enabled kinetic destruction of an adversary to attain unequivocal political outcomes. This stance has served the United States well over the centuries; but as American conventional proficiency continues to climb, so does the probability of asymmetric engagements—as illustrated in Afghanistan and Iraq today. The unwillingness of modern adversaries to oppose the United States in the conventional arena reduces the importance of violent destruction in a conventional sense while simultaneously elevating the primacy of contested ideas. This phenomenon suggests that the American proclivity to conduct military campaigns featuring conventional kinetic strikes may need reexamination. A harder look at the military's nonkinetic capabilities—founded in the timelessness of information and powered by the technologies of the information age—is in order.

The purpose of this paper is to determine whether or not America's kinetic emphasis represents an effective method to achieve the stated political objectives of the global war on terror (GWOT). Toward that end, the paper addresses four questions:

- What constitutes an effective military strategy?
- Does America exhibit a characteristic method of waging war?
- If so, what traits best depict that model?
- What model would best represent the antithesis of America's preferred form of warfare?

Methodology

This paper proposes a three-part framework to examine military effectiveness and to construct kinetic and nonkinetic models using the Hegelian notion of the dialectic. This framework is then used to analyze US military efforts in Operation Enduring Freedom (OEF) in both the kinetic and nonkinetic realms. Finally, this study evaluates current military strategy and suggests future areas of emphasis.

Framework for Effectiveness

Constructed upon a Clausewitzian foundation that assumes an immutable nature of war composed of reason, chance, and violence, a framework for examining military effectiveness is erected in three parts in chapter one. First, war's political object carries paramount importance as the guiding force impelling all military operations. The attainment of the political object as expressed by national policy objectives represents the only true measure of military effectiveness.

Second, war consists of both moral and physical realms, with each arena harboring access to potential sources of leverage known as centers of gravity (COG). By identifying correctly and influencing appropriately these COGs, protagonists may more readily attain the war's political object. This paper illuminates primarily adversary COGs, although occasional impacts to American domestic opinion are discussed. Adversary COGs are identified both from their linkages to stated political objectives as well as from expressed motivations for conflict. Although any given COG may theoretically be influenced by physical or psychological strategies, this paper examines military actions affecting each COG through either a kinetic or nonkinetic lens. For the purposes of this paper, Clausewitz's "moral" element of war is synonymous with the mental and psychological aspects of conflict, whether exemplified by adversary motivation and intention or popular support for a contested idea.

Third, any clash of opposing wills occurs along a continuum of escalating violence. Although both the moral and physical elements are present continually in war, each aspect and its requisite suite of compatible military capabilities assume greater significance depending upon the degree of violence within a particular conflict. As conflicts approach maximum levels of violence, or "total war" in the abstract sense, the physical realm and kinetic skills assume greater significance; conversely, at lesser degrees of violence, the psychological component and nonkinetic capabilities deserve elevated attention. Although this relationship may not be absolute, it serves well as a starting point for analysis. When examining particular actions within OEF, a temporal judgment of the conflict's position along Clausewitz's continuum of violence should suggest which array of capabilities military strategy should feature. Thus, accurate perceptions of the desired political end state, the relevant COGs, and a conflict's placement along Clausewitz's spectrum of violence emerge as the keys to a framework for determining the effectiveness of a strategic military strategy.

Kinetic/Nonkinetic Dialectic

Chapters two and three are crafted with the Hegelian notion of the dialectic in mind. Georg Wilhelm Friedrich Hegel envisioned the dialectic process as a descriptive exercise rather than a progression of deductive logic. In the abstract, the inadequacy of any given concept inspires a process of negation that generates an "anticoncept." These two diametrically opposed concepts struggle against one another to yield intuitively an entirely new conception. This phenomenon is described today by the triad of thesis-antithesis-synthesis, although Hegel never used these terms.[2] Furthermore, much like the view that portrays strategy as "a plan for attaining continuing advantage," the synthesized conception becomes a fresh thesis, and dialectic intuition through the process of negation begins anew.[3]

In this vein, chapter two constructs America's kinetic model to serve as the prevailing thesis of US military strategy. Founded upon the recurrent alternation of isolationist and interventionist foreign policies throughout US history, the kinetic model features a US military philosophy preferring the employment of disproportionate destructive force to limit the duration of armed conflict as well as to minimize friendly and foreign casualties. Furthermore, the United States possesses a messianic self-perception that yields a national tendency to exert an uncompromising stance regarding military endeavors while avoiding tactics of persuasion whenever possible. Finally, by subordinating intentionally military action to civilian leadership, the US system of government tends to insulate military actions from overarching political strategy as both politicians and generals are content to operate within their own spheres of expertise. Thus, coupled with the expected elements of firepower and maneuver within the physical realm, the kinetic model in this study also encompasses the US military tradition of acting somewhat divorced from political objectives while seeking to compel an adversary with destructive force. Significant military actions within OEF are analyzed through this lens to determine their congruency to both the stated political ends and the pertinent COGs.

Chapter three erects a nonkinetic model with which to examine OEF, with the understanding that an absolute antithesis of any concept cannot exist in reality. By outlining briefly the genesis of the pervasive discussion of an informational RMA within today's US military, the chapter begins by describing what the nonkinetic model is *not*: an information technology-enabled method to conduct kinetic operations more quickly and efficiently. After sifting through the jumbled taxonomy of information operations (IO), then removing kinetic elements and those that support primarily kinetic effects from the accepted elements of IO, a nonkinetic model featuring public affairs (PA), civil affairs (CA), and psychological operations (PSYOP) emerges. Furthermore, the nonkinetic model contains a public diplomacy component to lash this military collection of capabilities to overarching political objectives; this component ensures also that the nonkinetic model is as distinct as possible from its kinetic counterpart. The chapter concludes that a nonkinetic model emphasizing influence and persuasion in the moral arena represents the antithesis of the kinetic model. As with the kinetic model, significant military actions and policy statements within OEF are analyzed with this lens to determine their congruency to both the stated political ends and the pertinent COGs.

Application to Operation Enduring Freedom

Chapter four employs the three-part framework with the kinetic and nonkinetic models to evaluate the military effectiveness of OEF. Specifically, this chapter outlines the political end states and relevant COGs associated with the GWOT and then views US military operations in OEF from a kinetic and nonkinetic perspective to determine effects. Additionally, it analyzes OEF chronologically to highlight major shifts in military strategy or policy emphasis and to better illustrate the tension between the kinetic and nonkinetic perspectives. Finally, the paper offers a preliminary examination of Operation Iraqi Freedom

(OIF) utilizing the same methodology with cursory indications suggesting experiences similar to OEF.

The paper concludes with a discussion of an emergent synthesis of US military approaches. This synthesis of the kinetic and nonkinetic models incorporates elements of both but requires greater institutional acceptance and recognition of nonkinetic capabilities suited ideally to target the psychological dimension of war. Furthermore, anticipation of the degree of violence necessary within an impending conflict may guide planning efforts toward those collections of skills that best leverage the identified COGs. Finally, the realization that the level of violence required and the corresponding importance of either the physical or moral elements of war may shift over time in a given conflict is a key component to the synthesized conception of American warfare.

Limitations and Sources

This study's unclassified status represents its primary limitation. Particularly within the arena of nonkinetic activities, access to classified actions would undoubtedly uncover many additional operations for examination. A second limitation of the analysis concerns its reliance upon secondary sources. Given the breadth of this study and the time allotted for its completion, the ability to delve into the depths of American history, the origins of Islam, and the ever-changing operations of an ongoing war is restricted. However, using open-source national-level documents, executive-level policy directives, terrorist manifestos, congressional testimony, and the like, the study incorporates some degree of primary source material. Similarly, in the effort to limit the scope of this study—and despite the importance of protecting one's perceptions, information networks, and the electromagnetic spectrum from manipulation and exploitation—the study's nonkinetic model examines only the offensive aspects of IO. Regarding the nature of the extremist threat within the GWOT, this study examines only the Islamist movement associated with al-Qaeda.

Every effort is made to keep the analysis at the strategic level. Operational- and tactical-level action is presented only when direct linkages to strategic effects are apparent. Therefore, only the "major muscle movements" of US military strategy in OEF are examined, rather than specific engagements. Finally, the study considers action in OEF up to the time of this writing, while the section covering OIF offers only a cursory examination of its broad similarities with OEF. However, key differences exist between OIF and OEF that this study did not address, such as the perception of a terrorist-accessible weapons of mass destruction (WMD) program in Iraq, as well as the explosion of sectarian infighting following the fall of the Hussein regime. Therefore, before conclusions under this construct may be rendered for OIF, further study is required.

Notes

(All notes appear in shortened form. For full citation, see the appropriate entry in the bibliography.)

1. Strassler, *Landmark Thucydides*, 351–57.
2. Friedrich, *Philosophy of Hegel*, xl–xlii.
3. Dolman, *Pure Strategy*, 6.

Chapter 2

Clausewitzian Effectiveness

In war, the moral is to the physical as three to one.

—Napoléon Bonaparte

To judge the effectiveness of a particular military strategy, a framework to describe and evaluate a strategy's efficacy must be erected. Such a construction must consider not only the overarching nature of war but also war's principal constituent elements and their likely relationships. The comprehensive and timeless theory of war offered by the preeminent Prussian Carl von Clausewitz provides such a structure. After a brief justification of Clausewitz as the intellectual underpinning of this study's framework, this chapter presents Clausewitz's conception of the nature of war as comprised of reason, chance, and violence. It then establishes three components critical to military strategic effectiveness. First, a war's political object carries paramount importance as the guiding force impelling all military operations. Second, war consists of both moral (psychological) and physical realms, with each arena harboring potential sources of leverage known as centers of gravity. By identifying correctly and influencing appropriately these COGs, the war's political object can be more readily attained. Furthermore, indications suggest that the emergence of the information age may contribute to the increased importance of war's moral element. Finally, any clash of opposing wills occurs along a continuum of escalating violence. Using US Army colonel William Darley's interpretation of Clausewitz's spectrum of violence, this chapter asserts that although both the moral and physical elements are present continually in war, each aspect and its requisite suite of military capabilities assume greater significance depending upon the degree of violence within a particular conflict. Thus, accurate perceptions of the political end state, the relevant COGs, and a conflict's placement along Clausewitz's spectrum of violence emerge as the keys to determining the effectiveness of a military strategy.

In Defense of Clausewitz

The writings of Carl von Clausewitz, most prominently assembled in his posthumous 1832 classic *On War*, have been subjected to myriad misinterpretations and reinterpretations over the centuries. Early twentieth-century strategists perverted both Clausewitz's definition of war as "an act of force to compel our enemy to do our will" as well as his proposal that "direct annihilation of the enemy's forces must always be the dominant consideration."[1] Consequently, national mandates for horrific, force-on-force attritional stalemates on the grand scale of World War I resulted.

More recent scholars attack Clausewitzian theory in a calculated fashion, targeting specific features as weaknesses that render his theory irrelevant for the future. John Keegan cites war's inherent "irrationality" as proof of the fatal flaw in Clausewitz's seemingly reason-based theory of war, although Clausewitz's repeated references to war's tendency for irrational violence and its susceptibility to chance and uncertainty seem to account for this supposed shortfall.[2] Furthermore, Martin van Creveld suggests that "Clausewitz's ideas on war [are] wholly rooted in the fact that, ever since 1648, war ha[s] been waged overwhelmingly by states" and are therefore minimally applicable to the low-intensity conflicts of the modern era.[3] However, van Creveld is concerned mostly with ensuring that nations prepare for the correct form of war, a notion Clausewitz embraces early in his theory with his admonition "to establish . . . the kind of war on which [one is] embarking."[4] Positions such as these carry some modicum of accuracy when considered in isolation; however, when one digests the entirety of Clausewitz's theory, none retain their credibility.

David J. Lonsdale provides an outstanding defense of the relevance of Clausewitz for modern warfare in his work *The Nature of War in the Information Age*. By comparing the primary intellectual thrusts of Clausewitz, Antoine-Henri Jomini, and Sun Tzu across a variety of topics, Lonsdale concludes ultimately that Clausewitz's theory, albeit imperfect and wanting in certain areas, "still reflects the true nature of war most accurately." Furthermore, Lonsdale argues astutely that academia may bicker endlessly over Clausewitz's exact thoughts on particular issues; however, when one adopts a more general interpretation of Clausewitz's work, *On War* emerges as a comprehensive theory "for understanding almost any war regardless of its motivations."[5] This view coincides with Clausewitz's desire for his theoretical effort:

> Theory will have fulfilled its main task when it is used to analyze the constituent elements of war, to distinguish precisely what at first sight seems fused, to explain in full the properties of the means employed and to show their probable effects, to define clearly the nature of the ends in view, and to illuminate all phases of warfare in a thorough critical inquiry. Theory then becomes a guide to anyone who wants to learn about war from books; it will light his way, ease his progress, train his judgment, and help him to avoid pitfalls.[6]

Thus, by presenting the chief components of Clausewitz's theory of war in a general, holistic fashion, the theory's universal applicability to the determination of military strategic effectiveness should become apparent.

The Nature of War

Clausewitz contends that the collective decision to wield the instrument of war as an expression of policy should drive commanders to consider "the first . . . and the most comprehensive . . . of all strategic questions"; that is, "to establish . . . the kind of war on which they are embarking; neither mistaking it for, nor trying to turn it into, something that is alien to its nature." Successful navigation of this daunting task implies an understanding of the nature of war, a concept encompassed by Clausewitz's "paradoxical trinity." Thus, in Clausewitzian terms, war's nature is "composed of primordial violence . . . ; of the play of chance and probability . . . ; and of its element of subordination, as an instrument of policy, which makes it subject to reason alone." Although these three elements of violence, chance, and reason possess a variable interrelationship, a theory "that ignores any one of them or seeks to fix an arbitrary relationship between them would conflict with reality to such an extent [as to] be totally useless."[7] Therefore, each of these ideas and their contributions to a framework of strategic military effectiveness will be addressed in turn.

The preeminence of war as an expression of policy throughout Clausewitz's work demands that the element of reason instigates the discussion of war's nature. Clausewitz's oft-quoted dictum that "war is not merely an act of policy but a true political instrument, a continuation of political intercourse carried on with other means" suggests that war can only be the result of a thoughtful choice by decision makers possessing control of an entity's war-fighting apparatus. These civilian leaders translate this original motive for war into the war's political object, or in modern lexicon, the desired political objective or strategic end state. This political objective determines not only the "military objective to be reached" but also "the amount of effort it requires." Thus, despite contemporary assertions to the contrary, military operations should never be conducted separately or distinctly from civilian policy guidance, and "no major proposal required for war can be worked out in ignorance of political factors."[8] In other words, the initial, reasoned motive to pursue conflict establishes the political goal that all military action must support. Only the achievement of this political, strategic objective represents true success in the endeavor as lower-level tactical success "though beneficial, has little meaning if it cannot be translated into the attainment of policy objectives."[9] Nazi strategist and career general staffer Field Marshall Erich von Manstein expresses this observation clearly with his inability "to find any satisfaction" in the German tactical successes of 1941: "No one was clear any longer [about] what higher purpose all these battles were supposed to serve."[10] Thus, a perfectly executed war that does not culminate in the realization of the political objective is nonetheless a failure, and war without a rationale "is just mindless violence."[11] Finally, modern informational technologies may serve to shorten the linkages between the political, strategic objective and the military's tactical action through enhanced information exchange. Should such a "technologically-induced compression" occur, "the link of war to policy" outlined by Clausewitz only strengthens.[12] Therefore, a military strategy or operation can be considered effective only if it contributes to the strategic, political objectives established by national leaders, and this relationship is expected to continue into the future.

The Clausewitzian concept of chance within the nature of war attempts to account for two significant phenomena prevalent in every clash of arms. First, the assertion that "war is nothing but a duel on a larger scale" demands careful recognition of one's adversary as a calculating, reacting foe and that the entirety of armed conflict is a uniquely human affair. The decision to embark upon the path of war is rendered by humans, as is the choice to resist. This collision of two living entities on a grand scale also envelops the innumerable frailties of individual participants—the nameless, faceless ones who must combat their own fears as well as the enemy. As such, "absolute, so-called mathematical, factors never find a firm basis in military calculations," and "in the whole range of human activities, war most closely resembles a game of cards."[13] Finally, war often surrounds or abuts segments of humanity that may not actively participate in violence but may nonetheless play a pivotal role in securing strategic success for either side of the active belligerents. This centrality of humanity to the conduct of warfare lays the foundation for the exploitation of war's moral and psychological factors by its participants, a stratagem any effective military approach should consider.

Clausewitz's second component of chance addresses the notion of friction in warfare. Borrowed from mechanical engineering, Clausewitzian friction "more or less corresponds to the factors that distinguish real war from war on paper," and it includes the unpredictability associated with the foibles of humanity. Friction, however, is more expansive than the systemic degradation attributable to man's fallibility. The concept encompasses "countless minor incidents—the kind you can never really foresee—[that] combine to lower the general level of performance, so that one always falls far short of the intended goal."[14] Although continually present in war, friction may be reduced by sheer human determination, training, or technology, according to Clausewitz. From this, many proponents of modern informational warfare imply that friction may be eliminated completely through the efficient transmission of superior intelligence or enhanced battlespace awareness.[15] However, Barry Watts and others remind military strategists that "general friction will continue to be central to future warfare regardless of technological changes in the means of combat," and even *Joint Vision 2020* asserts that "information systems, processes, and operations add their own sources of friction and fog to the operational environment."[16] Thus, the imperfections of man, the ubiquity of friction, and the pervasiveness of each throughout warfare suggest that uncertainty and ambiguity will always exist in war. Therefore, an effective military strategy should not only anticipate these elements of Clausewitzian chance but also capitalize upon their existence by leveraging their links to the moral and psychological aspects of war.

The final facet of the Clausewitzian nature of war addresses the degree of violence inherent in warfare. Clausewitz establishes initially the characteristics of an abstract war, describing armed conflict as it would exist without restraint. At its core, Clausewitz contends, "war is an act of force, and there is no logical limit to the application of that force." Furthermore, "force—that is, physical force . . . is thus the means of war," and any belligerent that "uses force without compunction, undeterred by the bloodshed it involves . . . will gain the upper hand."[17] This action will drive the opponent to follow suit, thereby propelling each side into an ever-escalating spiral toward the extreme violence of "total war."

Despite this tendency of war to devolve into a maelstrom of violence for its own sake, Clausewitz is quick to highlight that this phenomenon does not exist in reality. Rather, because war does not erupt spontaneously and is the product of psychological and material calculations, neither side in a conflict will ever commit the entirety of its resources toward the venture.[18] Furthermore, physical impediments such as geography also inhibit war from reaching its theoretical potential for absolute violence, as does the existence of chance's human and material frictional elements. Thus, political objectives, born of a cost-benefit decision calculus that considers both moral and physical factors, serve not only to guide military action but also to limit it. Nonetheless, the violent, physical aspect of warfare can never be discounted, as even indirect, bloodless victories carry with them the deterrent effect of probable success in battle. This capability for tactical prowess demands an attention to the physical dimension of combat, an aspect of military planning that any effective strategy must feature.

Thus, Clausewitz's perception of the nature of war as composed of reason, chance, and violence contains a triumvirate of seeds necessary for an effective military strategy to germinate into political fruition. The essential requirement that the reasoned, political objective remains the singular focus of all military efforts must represent the centerpiece of a sound military strategy. The strategy must account also for the moral and psychological elements harbored by war's humanity, while simultaneously recognizing that friction and uncertainty can never be eradicated. Finally, constant acknowledgement of war's intrinsic violence will ensure that diligent attention devoted to the physical aspects of conflict yields tactical and operational success in support of strategic political aims.

Centers of Gravity

Clausewitz's discourse on war's nature touches upon the existence of both moral and physical realms within the conduct of warfare, but other sections of his work reveal this observation more explicitly. Clausewitz asserts that "essentially war is fighting. . . . Fighting, in turn, is a trial of moral and physical forces through the medium of the latter."[19] Here the physical aspect of conflict seems to command center stage when engaged in a collision of wills, while moral considerations, albeit worthy of attention, comprise only secondary importance. A strategist adopting ill-advisedly a strict interpretation of Clausewitz would seize upon this relationship and identify immediately the physical paths to attack to secure the destruction of the adversary's armed force. Any thought of military action to exploit psychological vulnerabilities within the moral arena would garner lesser support, if any at all. Surprisingly, this mentality does not diverge significantly from the military employment methodology currently enjoying favor within the US national defense community.

A more panoramic view of Clausewitz's appreciation for the moral element of force, however, reveals the complete perspective:

> Moral elements are among the most important in war. They constitute the spirit that permeates war as a whole, and at an early stage they establish a close affinity with the will that moves and leads the whole mass of force. . . . The effects of physical and psychological factors form an organic whole which, unlike a metal alloy, is inseparable by chemical processes. In formulating any rule concerning physical factors, the theorist must bear in mind the part that moral factors may play in it. . . . Hence most of the matters dealt with in this book are composed in equal

> parts physical and of moral causes and effects. One might say that the physical seem little more than the wooden hilt, while the moral factors are the precious metal, the real weapon, the finely-honed blade.[20]

Thus, Clausewitz grants the moral sphere of combat equal footing with its physical counterpart, implying that both facets demand equivalent treatment in the formulation of effective military strategy. This view is not unique to Clausewitz, adding to its credibility; the human characteristic of conflict "places psychological considerations at the heart of warfare," a position shared by Sun Tzu, Jomini, and even Clausewitzian critic Martin van Creveld.[21] Therefore, the shared significance of both the physical and moral elements within warfare implies that obtuse interpretations of Clausewitz claiming "destruction of the enemy forces . . . [as] always the superior, more effective means," do not grasp the entire picture.[22]

"The aim of warfare should be what its very concept implies—to defeat the enemy," asserts Clausewitz. However, the presence of physical and moral spheres within warfare, each requiring equitable consideration during the formulation of military strategy, suggests that either or both realms may possess key points of influence critical to achieving the political objective. Clausewitz establishes the likelihood of this probability with his contention that "in war many roads lead to success, and . . . they do not all involve the opponent's outright defeat." Furthermore, pure physical destruction is always "a means to some other end," and when this is no longer true, "the engagement . . . in itself . . . is of no value." Clausewitz also defines carefully enemy destruction as "such a condition that [the enemy] can no longer carry on the fight."[23] This precise terminology does not require the physical annihilation of an enemy's armed force, only the removal of the adversary's capability to resist by either physical or psychological means. Thus, sources of leverage exist in both the physical and moral arenas that one may exert pressure upon, by either physical or psychological means, to eliminate an adversary's ability to wage war and thereby attain political victory.

Clausewitz identifies these pressure points as *schwerpunkt*, or centers of gravity, and their character varies with the strengths and vulnerabilities of any given belligerent. Clausewitz defines the term as "the hub of all power and movement, on which everything depends. Th[is] is the point against which all our energies should be directed."[24] A misguided interpretation of Clausewitz would note the singularity of the reference as well as the Newtonian fact that an object can possess only one COG and conclude erroneously that war-fighting belligerents also harbor a single COG. By relaxing this constraint and recalling Clausewitz's recognition of multiple paths to victory, a more accurate picture of combatants with the potential to harbor multiple COGs comes to light. Not only may belligerents possess more than a single COG, each COG may also be influenced by either physical or psychological means. For example, a strategy to attack an enemy COG of leadership may comprise direct kinetic assault of the applicable individuals or seats of government, PSYOP to foment internal insurrection, or a combination of both.

Clausewitz's intent in positing the notion of a COG is simply to identify probable paths to pursue during war that will most likely culminate in the defeat of the enemy, chance and friction allowing. In his personal experience, physical conquest of the largest concentration of enemy combatants typically represented the most appropriate path to an adversary's capitulation. However, contemporary theorists expand the COG concept beyond the fielded forces to include the connections between governmental or military system elements. John Boyd first proposed "striking at those vulnerable, yet critical, tendons, connections, and activities that permit a larger system COG to exist" in order to disrupt an enemy's ability to orient and react to unwanted aggression.[25] John Warden expounded upon this concept to arrive at his concentric five rings that represent a self-professed "arrangement of the COGs you would find in any system . . . , their location within the five rings indicative of their probable importance to the system, to the enemy that you're trying to change."[26] Thus, in seeking enemy defeat, Warden contends his center ring of leadership often proves to be the most critical COG, while Clausewitz focuses upon an enemy's army, and Boyd concentrates upon the associations between them. Although all three positions emphasize different loci of leverage within an enemy system, all three acknowledge also the existence of multiple paths to enemy defeat that touch upon both the physical and moral arenas of warfare.

This subtle shifting of emphasis from seeking victory through predominately physical means in Clausewitz's era to methods favoring the moral aspect of war in Boyd's time may continue. A 1994 US Air Force report confirms the growing importance of war's psychological element by naming explicitly informational efforts as a new COG for the military.[27] Others observe "a shift in the COG away from traditional methods of force and means of combat toward non-traditional methods, including information."[28] Alan Beyerchen propels the argument further into the future contending that humanity is approaching "World War IV," an epoch of conflict dominated by social, rather than organizational or technological, pressures. Building upon Beyerchen's thesis, Maj Gen Robert Scales, US Army, retired, anticipates "a shift in classical COGs from the will of governments and armies to the perceptions of the populations." In this environment, he contends, cultural awareness and the manipulation of ideas become the focus of national instruments of power, with political victory "defined more in terms of capturing the psycho-cultural rather than the geographical high ground."[29] Regardless of the accuracy of these predictions concerning the character of tomorrow's wars, the psychological elements of human conflict will remain at least as relevant to effective military strategy as the physical paths to victory pursued in the past.

Thus, Clausewitz's insight into the physical and moral realms of warfare provides opportunities for military strategists to identify and exploit COGs within or overlapping each sphere in the pursuit of the political objective. Understanding that both aspects of warfare are ever-present and that "the kind of war on which they are embarking" may illuminate policy makers to focus upon certain COGs over others, an effective military strategy must nonetheless account for war's physical and psychological elements. Accurate perception of the relevant COGs in each arena, along with comprehension of how best to influence them in one's political favor, form the framework of an effective military strategy.

The Continuum of Violence

The final aspect of an effective military strategy founded in Clausewitzian principles stems from the idea that the degree of violence varies among conflicts, and, as such, so does the relative importance of the war's physical and moral COGs. Clausewitz's abstract "total war" has already been presented; this condition of "pure . . . enmity unleashed" represents the polar extreme of violence in war, a theoretical position unattainable in reality.[30] Clausewitz clarifies further this concept:

> The more powerful and inspiring the motives for war . . . the closer will war approach its abstract, the more important will be the destruction of the enemy, the more closely will the military aims and the political objects of war coincide, and the more military and less political will war appear to be. On the other hand, the less intense the motives, the less will the military element's natural tendency to violence coincide with political directives. As a result, war will be driven further from its natural course, the political object will be more and more at variance with the aim of ideal war, and the conflict will seem increasingly political in character.[31]

This description identifies the intensity of the initial political motive, which by definition must emerge from war's psychological or moral component, as the driver of the conflict's degree of violence. This phenomenon alludes further to the existence of a continuum of violence that features rampant destruction at one pole and pure nonviolent political competition at the other. Thus, the moral factor of force remains evident throughout the spectrum of violence as the instigator of the conflict itself, but its active influence and relative significance as an area for exploitation expands toward the purely political pole. Conversely, as war approaches the destructive extreme, the relative importance of physical manipulation of COGs begins to rise. This relationship does not represent a universal principle or analytical maxim. It merely illustrates one method to frame the investigation of conflict; this study tests the veracity of this claim.

Darley seizes upon this Clausewitzian continuum of violence and suggests that "a war of pure violence would be characterized by such unbridled use of kinetic instruments that other instruments of political conflict would be reduced to virtual irrelevance." Likewise, conflict at the political extreme of the spectrum would "be contested in a manner completely devoid of violence [and] would be characterized as totally ideological . . . decided exclusively by ideas, words, and symbols."[32] Darley concludes therefore that a military's kinetic arsenal aligns well with the prosecution of predominately violent wars while its IO capabilities suit conflicts toward the political pole. His key observation states that "at their core, IO elements dealing with public information are political activities—the emphasis on psychological and influence aspects of political conflict become progressively more pronounced and dominant as the conflict becomes less dependent on violence."[33] From this, one may extrapolate that a military's kinetic weaponry supports ideally the physical exploitation of COGs en route to strategic victory while its nonkinetic instruments mesh well with their moral elements. Furthermore, Darley stresses the realization that "a vaguely defined threshold somewhere in the middle of the continuum" must exist, "the crossing of which signals a seminal change in the relationship between IO and kinetic operations."[34] At this juncture, either a physical, kinetic military operation or a psychological, nonkinetic campaign should become the supported military effort, depending upon the perceived location of the conflict along the continuum of violence. The supported effort enjoys primacy in planning the major thrusts of a given campaign and first call upon certain low-density/high-demand assets; however, both kinetic and nonkinetic activities continue regardless of their relative priority. Figure 1 offers a graphic depiction of this idea. Regardless of how the military strategy chooses to exploit the relevant collection of COGs, opportunities to address points of leverage in both the moral and physical spheres remain throughout the range of conflict.

Finally, Darley attempts to account for time in his model as a measure of the conflict's intensity while he places past conflicts statically along Clausewitz's continuum. This construct does not reflect the fluidity of actual combat and the chimerical nature of a nation's political will. It is important to note that the character of a given war may change over the course of the conflict, causing the appropriate military emphasis to shift during a single operation. Thus, a conflict's penchant for violence, as determined primarily by the magnitude of its political motive, may serve to guide military strategy in its efforts to focus upon the appropriate COGs by tapping either physical or psychological means. By choosing to emphasize the suite of capabilities that best engages the identified paths to political success, a military strategy elevates its effectiveness.

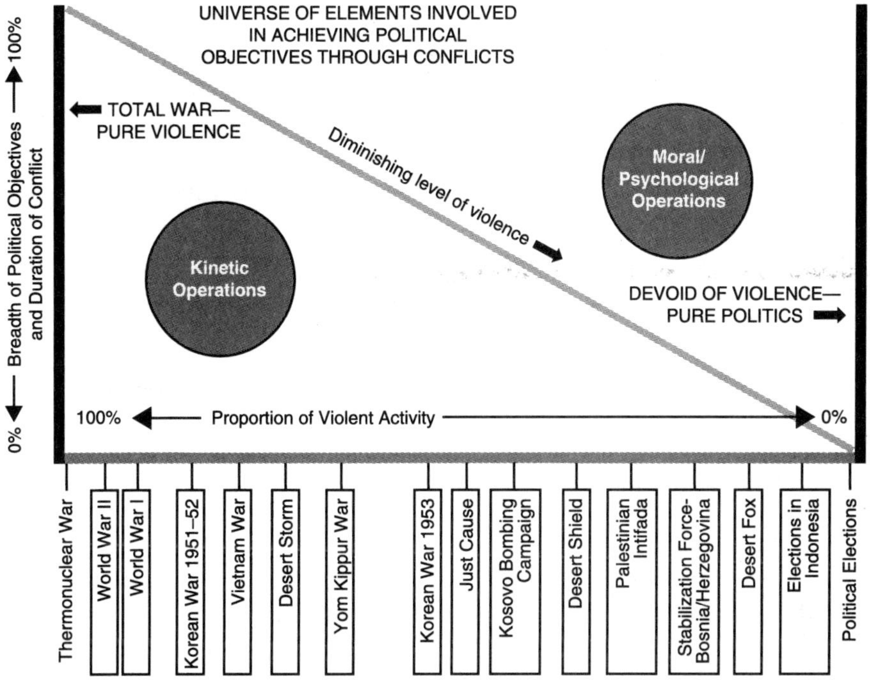

Figure 1. The continuum of violence. (Reprinted from Col William M. Darley, briefing, US Army Combined Arms Center, Fort Leavenworth, KS, subject: Information Operations Theory and Public Affairs, April 2004, slide 61.)

Summary

This chapter constructs a framework of strategic military effectiveness by expanding upon the fundamental ideas of Clausewitz. Despite assertions to the contrary, his theory of war remains the most comprehensive; considering the resurgence of strategic difficulties in modern war, it is also the most relevant. The centrality of war's subservience to its political objective throughout Clausewitz's theory provides military strategists with the desired end state toward which all operations must strive. War's ineluctable human element ensures that moral and psychological considerations will forever permeate armed conflict, and war's propensity toward destructive violence also preserves physical prowess among its perennial features. Therefore, both the moral and physical components of war harbor potential loci for adversarial influence—known as COGs—that an effective military strategy may target by either kinetic or nonkinetic methods. By recognizing the primacy of either physical or psychological tactics in the quest for political victory, the appropriate repertoire of military capabilities may be matched to the relevant COGs and receive subsequent operational emphasis. In such a fashion might a military strategy attain its political, strategic objective—the only measure of true effectiveness.

A final point concerning the termination of war demands elucidation. Clausewitz erects the political object as the desired strategic end state of armed conflict. This chapter built upon this fundamental principle to offer the attainment of the political objective as the aim of all military operations and the validation of an effective military strategy. However, achieving an "end state" implies the culmination of a series of events and a terminus of willful confrontation. This perception of a static strategic victory is not complete and does not portray an accurate depiction of the role of strategy. Everett Dolman contends that "continuation is the goal of strategy—not culmination. Strategy is . . . an unending process that can never lead to conclusion," and as such, strategy should "influence states' discourse in such a way that it will go forward on favorable terms."[35] Dolman's masterfully simple definition of strategy as "a plan for attaining continuing advantage" aligns perfectly with Clausewitz's view of war termination.[36] Clausewitz agrees that "the ultimate outcome of a war is not always to be regarded as final. The defeated state often considers the outcome merely as a transitory evil, for which a remedy may still be found in political conditions at some later date."[37] Colin Gray expounds further that "war should be waged not for the goal of victory, necessary though that usually is, but rather for the securing of an advantageous peace."[38] Therefore, not only must an effective military strategy achieve the political objective as outlined by a nation's civilian policy makers, it must also consider the long-term implications of military action to the nation's continued relationships with both the adversary and the world.

Notes

1. Clausewitz, *On War*, 75, 228.
2. Keegan, *History of Warfare*, 3.

3. Van Creveld, *Transformation of War*, 41.
4. Clausewitz, *On War*, 89.
5. Lonsdale, *Nature of War*, 26, 231.
6. Clausewitz, *On War*, 141.
7. Ibid., 88–89.
8. Ibid., 81, 87, 608.
9. Lonsdale, *Nature of War*, 79.
10. Manstein, *Lost Victories*, 202.
11. Lonsdale, *Nature of War*, 28.
12. Jablonsky, "US Military Doctrine," 36.
13. Clausewitz, *On War*, 75, 77, 86.
14. Ibid., 119.
15. This school of thought is addressed in greater detail in chap. 4.
16. Watts, "Clausewitzian Friction," 131; and *Joint Vision 2020*, 9–10.
17. Clausewitz, *On War*, 75–77.
18. Ibid., 79.
19. Ibid., 127.
20. Ibid., 184–85.
21. Lonsdale, *Nature of War*, 37.
22. Clausewitz, *On War*, 97.
23. Ibid., 90, 94, 96, 595.
24. Ibid., 595–96.
25. Boyd, "Patterns of Conflict," slide 41.
26. Kagan, *Finding the Target*, 121.
27. Coffman, *Strategic Environmental Assessment*.
28. Thomas, "Deterring Information Warfare," 81.
29. Scales, "Clausewitz and World War IV," 18.
30. Clausewitz, *On War*, 605.
31. Ibid., 87–88.
32. Darley, "Clausewitz's Theory," 75–76.
33. Darley, "Information Operations," slide 54.
34. Darley, "Clausewitz's Theory," 78.
35. Dolman, *Pure Strategy*, 4, 6.
36. Ibid., 6.
37. Clausewitz, *On War*, 80.
38. Gray, "How Has War Changed?" 18.

Chapter 3

The Kinetic Model and Its Prevalence

Sometimes one is moved to the despairing conclusion that Clausewitz wrote in vain, for all the influence he has had on the American way of war.

—Colin S. Gray

This chapter constructs a kinetic model of military operations based upon the characteristics of contemporary and past American force employments. By examining thematic trends throughout US history, this chapter concludes that a kinetic model exemplifying the American preferred method of waging war features short, destructive conflicts that seek to compel adversaries while often remaining separated from political objectives.

US history displays recurrent tension between foreign policies of isolationism and engagement. This oscillation contributes to a prevailing US military philosophy that prefers the employment of disproportionate destructive force to limit the duration of armed conflict as well as to minimize friendly and foreign casualties. The United States possesses also a messianic self-perception that, when coupled with its relatively short history of military successes, yields a national tendency to adopt an uncompromising stance regarding military endeavors. Finally, by subordinating military action to civilian leadership, the United States embraces a governmental system that tends to insulate military actions from overarching political strategy.

Destruction

America's foreign policy has oscillated between an inward and outward focus since its genesis in the late eighteenth century.[1] Born of a rising frustration with the politics of Europe, America's very status as a nation finds its foundations in a core desire to remain distant from international affairs. From Pres. George Washington's 17 September 1796 farewell address that outlined a foreign policy of "steer[ing] clear of permanent alliances" to Pres. Thomas Jefferson's memorable 1801 inaugural address vowing "entangling alliances with none," the fledgling, formative years of America's foreign policy favored an isolationist stance.[2] Reinforced by fortuitous geography and benign borders, America's inward focus emerged as the preferred foreign policy position of a nation content with the pursuit of individual freedoms and the accumulation of wealth. Despite its preferences, America discovered quickly that global economic wealth and national security cannot be achieved with pure political neutrality.

The sacking of Washington in 1814 propelled the United States to consider another venue to security—"enlarging, rather than . . . contracting its sphere of responsibilities."[3] Articulating the foundations of preemption and unilateralism that permeate contemporary US foreign policy, men such as John Quincy Adams advocated an expansion of US influence abroad to attain hegemonic status. Whether driven by economic, social, or purely security considerations, America embroiled itself repeatedly in international conflicts over the next 200 years. However, these expressions of global engagement culminated invariably in periods of military drawdown and a return to the comfort of isolationism. Thus, "for most of its existence, [the United States] has generally maintained a small and isolated armed force during peacetime, mobilizing for war and demobilizing quickly after it was over."[4] Therefore, America's historical use of its military instrument of power may be described as reluctant and governed by a national desire to return to an inward focus at the earliest opportunity.

America's cultural mind-set of historical reluctance to engage militarily in world affairs shapes the character of its armed endeavors in two fundamental ways. First, the United States favors the introduction of overwhelming force to ensure rapid, decisive military victory. Revealed compellingly in Russell F. Weigley's classic *The American Way of War*, the United States has learned throughout its history that a strategy of annihilation best limits the duration of armed conflict.[5] Only when resources cannot support such a strategy, or they are diverted in support of graver priorities, has the United States followed a military policy of attrition. The American experience in World

War I—where defeating large conventional armies required overwhelming masses of men and material—reinforced this notion. The purely destructive purpose of the US military surfaced within US Army doctrine during the interwar years with *Field Service Regulations* (1923) and Field Manual 100-5, *Tentative Field Service Regulations: Operations* (1939), describing "the ultimate objective of all military operations [as] the destruction of the enemy's armed forces by battle. Decisive defeat in battle breaks the enemy's will to war."[6] This view of unquestioned military victory as the quickest path to political success persists to this day.

Second, and stemming from this idea of achieving a rapid, crushing victory, America displays a stark aversion to casualties, particularly under certain circumstances. Given America's Judeo-Christian heritage that holds the sanctity of life dear, the United States seeks to circumvent unnecessary loss of life; furthermore, when "American society . . . does not care about the issues in dispute or it realizes that Washington is not seriously seeking decisive victory" this casualty aversion is even more pronounced.[7] It is important to note that America does accept the necessity of casualties under special conditions. Unambiguous national objectives that are "morally and politically lofty, . . . vital to the US," and relevant to "the interests of most Americans" elevate the country's tolerance to absorb casualties.[8] This fact has prompted some scholars to declare the American casualty-aversion assumption as myth; however, contemporary US military involvement often does not portray such explicit links to justifiable objectives, and popular outcries against baseless violence persist.[9] Thus, America, and by extension its military, remains vulnerable to an aversion of both foreign and indigenous casualties. It continues to envision war as "a short and sharp engagement" with "the purpose of American arms . . . to rout the enemy and then get out."[10]

Compellence

The United States historically displays the tendency to participate in military confrontations only reluctantly. However, when it does elect to exercise its military instrument of power, the nation seldom exhibits willingness to compromise. Instead, the United States seeks to compel adversaries to comply with its aims. Compellence represents the offensive element of coercion, comprised of the administration of punishment "*until* the [adversary] acts, rather than *if* he acts" (emphasis in original).[11] Thus, once the United States chooses to unleash the destructive force of its military, that punitive action ceases only upon enemy compliance to US demands. Furthermore, such an uncompromising stance coupled with decisive, destructive victory places a nation "in a position to command total obedience on the part of the defeated adversary"; this enviable arrangement "open[s] the way to the unhindered realization of your political objectives, whatever they might be."[12] As with the nation's fascination with isolationism, this US propensity to neglect negotiation during its military ventures also finds its underpinnings during the nation's revolutionary birth.[13]

Largely motivated by disgust with European tyranny, the initial immigrants to America found themselves confronted with the daunting task of conquering an unknown, hostile land. "The fact that the early Americans fought and succeeded against great odds to form the modern state system's first practicing democracy," coupled with their prevailing founding principles of personal and religious freedoms, engenders a national atmosphere of divinely directed destiny.[14] Early Puritan settlers viewed themselves as "chosen people" to fulfill a divinely directed terrestrial mission.[15] Thomas Paine's *Common Sense* embodies this messianic self-perception in its description of the United States as a "project for mankind," as does Jefferson's Declaration of Independence affirming the nation's "reliance on the protection of Divine Providence."[16] This national persona often translates into "a crusading zeal that wants to 'save' the world from itself in the name and image of . . . America," as in Senator Albert Beveridge's 1900 advocacy for imperialist expansion based upon "God's hand . . . in . . . the movement of the American people toward the mastery of the world."[17] Likewise, Pres. Woodrow Wilson's citing of "the principles that gave her birth and happiness" and "God's help" as America's justification for entry into World War I, as well as his League of Nations ratification plea that "nothing less depends upon this decision, nothing less than the liberation and salvation of the world," illustrate this American messianic worldview.[18]

Yale University professor John Lewis Gaddis contends Pres. George W. Bush today exudes a similar philosophy, calling for an "expansion of the empire of liberty" in response to 9/11 and a corresponding extension of democracy "everywhere."[19] Furthermore, and based upon its self-perception as a chosen nation, America assumes "that the political system of a place like Minnesota [will] also work well in Burma or Kenya or Cuba or Algeria."[20] This self-view of moral—and by extension political—superiority rarely allows, or even sees the need for, the consideration of alternate views during conflict.

America's relatively brief history and record of comparative military success further contributes to its employment of military might in an uncompromising fashion. Punctuated by a successful founding revolution, a civil war that managed to preserve the Union, "a series of westward expansions that encompassed the North American continent," and two global wars that made "the world safe for democracy," America's brief military experience has been one of memorable accomplishments.[21] These triumphs not only embellish America's self-perceptions of superiority, they also embolden military action that disregards considerations of compromise. Although the US experience in Vietnam looms large as an obvious exception to America's impressive military record of success, the fact that the nation's ignominious withdrawal proved so traumatic suggests an American expectation for military victory. Thus, stemming from its many military successes within a short institutional memory, the United States exhibits an explicit preference for the adversary's complete submission and unconditional surrender in its military conflicts.

Finally, and related to America's self-image as a role model for the world, the United States' traditional military focus of unyielding resolve contributes to a national culture that considers efforts to influence foreign perception as immoral. Given the "attractiveness of the institutions and ideas that compose the American system," along with the impressive effects of America's "soft power," some contend there exists "no need to overtly persuade" foreign audiences.[22] Consequently, the United States needs only to lead by example, and the world will not only "recognize how wonderful we are" but also strive "to emulate us."[23] Although this position has waned in recent years, the notion still persists, coloring military efforts to influence an adversary psychologically with an aura of trickery and underhandedness. Democracies in general and the United States in particular "have often abstained from using PSYOP, employing [them] only reluctantly during conventional wars, and dismantling the . . . painstakingly-assembled . . . PSYOP organization . . . in a matter of weeks, once the war is over."[24] Thus, America's messianic self-perception, coupled with its brief record of generally successful military interventions, inspires a military stance of unwavering compellence toward its adversaries that regards psychological operations as not only un-American but also unnecessary.

Distinct from Politics?

Related to America's compellent and destructive stance in military efforts, an attitude that military action represents an alternative to political discourse rather than its extension pervades US strategic thought. Contrary to the Clausewitzian dictum that describes war as a "political instrument, a continuation of political intercourse, carried on with other means," the Western way of war often operates separately and distinctly from political interference.[25] Stemming from their "high premium on individualism," Western militaries "are often subject to criticism and civilian complaint" based upon their desire and intent to operate independently under the direction of unchanging, unambiguous political objectives.[26] Furthermore, Victor Davis Hanson describes the Western view of war as "a method of doing what politics cannot," a position that remains noticeable in the US approach to warfare.[27] Thus, as Antulio J. Echevarria, the current director of research at the Strategic Studies Institute, contends, America considers war "an *alternative* to bargaining," more indicative of a "way of *battle* than an actual way of war" (emphasis in original). Consequently, the US military considers rarely the "gritty work of turning military victory into strategic success" and instead exhibits a fascination with the precise refinement of war's Clausewitzian grammar rather than its overarching logic.[28]

The United States intentionally divides its political and military realms by subordinating military command to civilian leadership. This bifurcation of US strategic thought yields "two separate spheres of responsibility, one for diplomacy and one for combat."[29] As a result, the US military tends to concentrate on winning battles and campaigns while policy makers focus on the diplomatic struggles that precede violence. Samuel Huntington advocates such a civil-military relationship as ideal to both areas of expertise in his seminal work *The Soldier and the State.* His notion of "objective civilian control" seeks to maximize military professionalism by keeping its membership distinct and uninvolved in political matters.[30] This relationship finds expression, much of it positive, within the modern US military. The US Marine Corps' *Small Wars Manual* describes conventional war as violence used as a last resort "after diplomacy had failed. . . . Political leaders handed over to military men the problem that diplomacy had not solved and told them to deal with it."[31] Similarly, the "Code of the Army Officer" includes the "Tradition of Avoiding Matters of Politics" as a key tenet, and Gen William Westmoreland concludes that "when does a professional military man put his fingers into the political mud and try to influence the political mechanisms by his [direct or indirect] actions?—he doesn't."[32]

This American tendency to strive for political-military separation in the execution of its foreign policy may be rooted in its affinity for civilian militia as its armed instrument of power. Historian Walter Millis attributes America's oscillation between isolationist and interventionist foreign policies to the constitutional limitation of Army appropriations to only two-year increments.[33] Favoring popular liberty over a standing armed force, America's founders relied upon an amateur fighting force culled from its general population in times of emergency. This requirement of militia forces not only ensured popular support of an impending conflict, it also engendered the national desire to demobilize at the conclusion of hostilities and served to keep military matters separate from political discourse. Although the demands of modern security threats necessitated the construction of a standing army to maintain proficiency and professionalism, efforts such as Gen Creighton Abrams's organizational shift of critical Army capabilities to the Reserve force in the 1970s reflect this continued national preference for the militia concept and its associated distinction from politics.[34]

Although this characteristic of American war fighting may represent an acceptable price to preserve civilian control of the military, it disregards nonetheless Clausewitz's presentation of policy and war as a logical continuum. This conscious, governmental decision to subordinate military action to civilian control comes often at the expense of a military strategy unified with a potentially dynamic political purpose. Thus, despite Clausewitz's admonition to preserve the national political objective as the unitary focus of one's military endeavors, the United States employs its military instrument to indicate the termination of political dialogue rather than its extension.

Contemporary Illustrations

Evidence that America displays an unrelenting posture to conduct short, overwhelming, destructive conflicts that minimize casualties and that operate somewhat separately from political discourse remains prevalent in both the composition and recent employment of the US military. That the Vietnam experience came to typify how military operations should not be accomplished—along with the rejection of the conflict's piecemeal, gradualist approach to military strategy—is not surprising. As the military struggled to recover from the debacle of Vietnam, many advocated a return to proven principles, and in many respects the adoption of the Weinberger-Powell doctrine and its corresponding emphasis upon overwhelming, destructive force represented such "a return to traditional American ways."[35] This destructive vision manifested itself in America's one-sided victory over Iraq in Operation Desert Storm (ODS), a conflict that also featured physical expressions of US society's thoughts on war.

As Clausewitz observes, "Very few of the new manifestations in war can be ascribed to new inventions or new departures in ideas. They result mainly from the transformation of society . . . and new social conditions."[36] ODS portrayed not only the resurgence of America's desire to dominate physically but also to minimize casualties. It is in this conflict that America's precision weaponry reached maturity, and the nation began to recognize "surgical strikes" as a legitimate military option to minimize collateral damage, whether human or material. Standoff weapons proved effective at minimizing risks to friendly troops, and ground-aided precision strikes from airborne bombers further refined this capability.[37] This renaissance of American destructive force combined with casualty avoidance has come to embody a so-called "new American way of war." Some describe the "new" US military position on war as seeking "a quick resolution using . . . speed and knowledge, cutting-edge . . . forces, and precision strikes against key targets."[38] Others cite "network-centric warfare, rapid decisive operations, and shock and awe" as characteristic of current US military thought, all seeking to "take down an opponent quickly."[39] Prevailing Defense Department "transformation" efforts all feature "speed, jointness, knowledge, and precision," and the burgeoning budgets favoring a two- to seven-fold increase in expenditures toward nonlethal weaponry over the next few years suggest a continuing interest in keeping casualties low.[40] Thus, current "transformational" efforts seem only to "enable the US armed forces to do better what they already do well."[41] Although modern technology cloaks these concepts in a veil of innovation, they represent in reality a return to the foundations of American military thought—that is, a singular focus on defeating an enemy decisively in battle, while minimizing casualties, and with little regard for the pursuit of broader political aims.

Expressions of America's idyllic self-perception still abound in the nation's political rhetoric, adding further resolve to America's stance of military compellence. Pres. Ronald Reagan repeatedly referred to the United States as a "city" or "shining house" on a hill, quoting John Winthrop's 1630 rendition of nascent America to suggest the nation represented a "beacon offering hope to other nations and peoples."[42] Pres. George W. Bush cited Abraham Lincoln's second inaugural address in his national call to combat terrorism by calling America the world's "last, best hope," and his declaration of a "crusade" against the same threat cemented further the US military's inability to compromise.[43] These contemporary executive expressions of America's historic self-image illustrate the relevance of an attitude of compellence to the current employment of the US military.

Recent military success has also intensified America's sensation of martial superiority. From ODS to "Somali clansmen, Bosnian Serbs, Serbs again, hapless Talibans, [and] Iraqis in a return engagement," the US military has largely dominated its foes, at least in the physical, kinetic realm.[44] These repeated demonstrations of military prowess against a series of inferior opponents serve to mask any underlying strategic deficiencies and can only add to the nation's belief in the efficacy of its current, unyielding posture. Furthermore, these sequential successes may also encourage the growing frequency of US military interventions. Without a balancing peer power to require the husbanding of military capabilities, along with the reduced risk of US casualties, "wars for less-than-vital interests [become] more palatable" to US society.[45] This recent emergence of "wars of discretion, rather than necessity" suggests that America's foreign policy pendulum has swung back toward the engagement pole and that the nation is more willing to demonstrate its perceived role as a global crusader.[46] Thus, America's comfort in its own military dominance, together with its continued perception of itself as the world's shining example of an ideal society, helps to propagate its military preference for compellence.

The United States continues to display a tendency to conduct military operations in political isolation as well. Besides its emphases on destruction and compellence, the aforementioned Weinberger-Powell doctrine also implies military action unencumbered by political interference once the decision to unleash punitive destruction is transmitted to military authorities. This philosophy of political leadership standing aside once the military propels itself upon the enemy continues to channel military efforts on the battle while politicians wait for the dust to settle before reengaging in serious diplomatic dialogue or entertaining thoughts of post-conflict operations. Recent US forays into Afghanistan and Iraq reveal this political-military disintegration as well by "placing more emphasis on destroying enemy forces than securing population centers and critical infrastructure and maintaining order" following hostilities.[47]

Finally, a recent confrontation between Senator John Warner and newly-appointed commander of the Multi-National Force in Iraq, Lt Gen David Petraeus, illustrates America's attempts to maintain clear divisions between its political and military strategies. When asked by Senator Joseph Lieberman if "a Senate-passed resolution of disapproval for [President Bush's] new strategy in Iraq [advocating the infusion of an additional 21,500 troops] would give the enemy some encouragement, some feeling that . . . the American people [were] divided," General Petraeus responded, "That's correct, sir." Senator Warner then chastised the general for commenting obliquely on the wisdom of the Senate's proposed resolution and advised Petraeus to "examine the transcript" to identify certain responses that he "might later regret" as suggestive of painting the Senate as "aiding and abetting the enemy."[48] Pundits interpreted this exchange as America's political arm reminding its military implement to "stay in its own lane" rather than comment upon the political motivations behind decisions affecting military matters. Thus, America's bifurcation of its grand strategy into distinct political and military elements continues into its most recent foreign entanglements.

Summary

Thus, beyond the expected elements of "conventional, force-on-force" confrontations, an emphasis upon "dropping bombs, shooting bullets and artillery rounds," and operations centered upon "troop movements, maneuvers, . . . and the occupation and control of physical terrain," the kinetic model of this study also encompasses the US military tradition of acting somewhat divorced from political objectives while seeking to compel an adversary with destructive force.[49] Included within this construct are America's general aversions to casualties as well as psychological operations.

This kinetic model will be applied to OEF to determine the congruency of its military means to both the stated political ends and the pertinent centers of gravity. To account for the fact that "there is more to war than warfare," a nonkinetic model will also be applied for comparative purposes.[50]

Notes

1. Quester, *American Foreign Policy*, 1.
2. Snow and Brown, *United States Foreign Policy*, 12.
3. Gaddis, *Surprise, Security, and the American Experience*, 12–13.
4. Snow and Brown, *United States Foreign Policy*, 13.
5. Weigley, *American Way of War*, xxii.

6. US War Department, *Field Service Regulations*, 77; and US War Department, Field Manual (FM) 100-5, *Tentative Field Service Regulations: Operations*.
7. Gray, "How Has War Changed?" 20.
8. Snow and Brown, *United States Foreign Policy*, 258.
9. Gray, "How Has War Changed?" 26.
10. Hendrickson, "Toward Universal Empire," 9.
11. Schelling, *Arms and Influence*, 70.
12. Kennan, *Memoirs: 1925–1950*, 308.
13. Scholars such as Thomas A. Bailey contend that negotiation and arbitration represent fundamental attributes of historical US foreign policy. However, these traits are most evident in advance of hostilities. See Bailey, *Diplomatic History of the American People*.
14. Snow and Brown, *United States Foreign Policy*, 10.
15. Judis, *Folly of Empire*, 14.
16. Ryan, *US Foreign Policy in World History*, 22.
17. Judis, *Folly of Empire*, 4.
18. Snow and Brown, *United States Foreign Policy*, 14–16; and Judis, *Folly of Empire*, 14.
19. Gaddis, *Surprise, Security, and the American Experience*, 109–10.
20. Quester, *American Foreign Policy*, 56–57.
21. Snow and Brown, *United States Foreign Policy*, 13–14.
22. Nye, *Soft Power*, 11–15; and Gough, *Evolution of Strategic Influence*, 2.
23. Gough, *Evolution of Strategic Influence*, 2.
24. Schleifer, "Democracies, Limited War," 50.
25. Clausewitz, *On War*, 87.
26. Hanson, *Carnage and Culture*, 22.
27. Ibid.
28. Echevarria, *Toward an American Way of War*, 1.
29. Ibid., 7.
30. Huntington, *Soldier and the State*, 83–85.
31. US Marine Corps, *Small Wars Manual*, ix.
32. Nagl, *Learning to Eat Soup with a Knife*, 201.
33. Millis, *Arms and Men*, 48.
34. Kitfield, *Prodigal Soldiers*, 199.
35. Joes, *Saving Democracies*, 4.
36. Clausewitz, *On War*, 515.
37. Theisen, *Ground-Aided Precision Strike*, 1–2.
38. Metz, *Armed Conflict in the 21st Century*, 87.
39. Echevarria, *Toward an American Way of War*, 16.
40. Ibid.; and Allison and Kelley, *Nonlethal Weapons and Capabilities*, 2.
41. Gray, "How Has War Changed?" 21.
42. Snow and Brown, *United States Foreign Policy*, 10–11; and Kilroy, "Public Diplomacy," 137.
43. Kilroy, "Public Diplomacy," 138.
44. Gray, "How Has War Changed?" 18.
45. Echevarria, *Toward an American Way of War*, 9.

46. Gray, "How Has War Changed?" 20.
47. Echevarria, *Toward an American Way of War*, 13.
48. Ricks, "Leave Politics to Us," A21.
49. Hall, *Stray Voltage*, 5.
50. Gray, "How Has War Changed?" 21.

Chapter 4

The Nonkinetic Model

The material focus of war, which tends to be the focus of attention, is less important than its social, cultural, and political contexts and enablers.

—Jeremy Black

The exponential increase of speed and capability evident in modern information technologies has spurred widespread discussion surrounding the possible emergence of a global information revolution. This phenomenon has penetrated strategic military thought by spawning similar debate concerning the prevalence of a contemporary revolution in military affairs (RMA). The novelty of these informational capabilities in the conduct of warfare, coupled with intense interservice scrambling to lay claim to the promising mission areas, has produced a confusing collection of conflicting and overlapping definitions. This chapter outlines briefly the genesis of the pervasive discussion of an informational RMA within today's US military, sifts through and organizes the subject's jumbled taxonomy, and arrives at a nonkinetic model that is as distinct as possible from its kinetic counterpart. The nonkinetic model emerges from a process of disaggregating the accepted panoply of IO and then removing kinetic elements and those that support primarily kinetic effects. Reassembling the remaining components of IO builds a structure of nonkinetic capabilities focused upon IO's core purpose, featuring public affairs, civil affairs, and PSYOP. Together these elements represent the core of this study's nonkinetic model. Furthermore, to lash this collection of military capabilities to overarching political objectives, the nonkinetic model incorporates a public diplomacy component. Thus, the chapter concludes that a nonkinetic model emphasizing the influence and persuasion of an adversary's mind-set through the means of PA, CA, PSYOP, and public diplomacy represents the antithesis of the kinetic model.

An Informational Revolution in Military Affairs?

Impressive advances in computer and informational technologies compel many to extrapolate an exponential increase in the importance of information and knowledge to future human interaction in general and warfare in particular. In 1965, Intel Corporation chairman Gordon Moore observed that "the surface area of a transistor . . . reduce[s] by approximately 50 percent every twelve months."[1] This relationship was revised to 18 to 24 months in 1975, and since then, the ability to double the quantity of components on an integrated computer chip has translated into a corresponding doubling of processing speed about every two years. Moore's Law contributes to the contemporary explosion of information quantity as well, which also exhibits a pattern of doubling every two years. Most recently, 2003's 57 billion gigabytes of global information exchange burgeoned to over 100 billion gigabytes in 2005, which is "roughly equivalent to two trillion four-drawer filing cabinets of hard-copy documents."[2]

Such mind-numbing figures induce thoughtful scholars to proclaim an information revolution that represents the emergence of a third, informational "wave" of global civilization out of the agrarian- and industrial-based swells of earlier times.[3] This perceived global shift from machine- and labor-based industrial societies toward those favoring information and knowledge promises to some a new way of thinking about war, a vision of conflict that is less fraught with uncertainty, less violent, and less frequent.[4] Should the dawn of an information age truly affect warfare in the manner some suggest, then "supremacy in handling information" will vault to the forefront of America's national security concerns, and the world may truly be embroiled in an RMA.[5]

Born of the Soviet concept of a military technical revolution, definitions of an RMA in today's national security lexicon vary in their views of the nature of war, but most emphasize technology. At the more controversial frontier, the Center for Strategic and International Studies (CSIS) perceives an RMA as "a fundamental advance in technology, doctrine, or organization that renders existing methods of conducting warfare obsolete."[6]

Such an extreme statement borders on claiming that an RMA represents an alteration of the very nature of war, an impossible phenomenon according to Clausewitz. Unlike the CSIS version, David Mets's more moderate interpretation retains the perspective of several millennia of warfare history and describes an RMA as "a rapid change in military technology, doctrine, and organization leading to a sweeping new way that wars are fought."[7] This version aligns more closely with the Clausewitzian notion that war's character, but not its overarching nature, may change to reflect advances in technology or tactics. Regardless, contemporary debate of this topic, particularly the technological aspects of a possible informational RMA, permeates military strategic thought and tickles the imagination of the national defense community.

The idea of an informational RMA exudes seductive allure to American society and, by extension, its military. Claims that an information revolution may increase US capacity to "collect vast quantities of precise data; to convert that data into intelligible information; to rapidly and accurately transmit this large quantity of information; to convert this information through responsive, flexible processing into near-complete situational awareness; and, at the limit, to allow accurate predictions of the implications of decision[s] . . . or actions" cause some to proclaim the impending power "to lift the fog of war."[8] The best example of the military's manifestation of this concept falls under the guise of network-centric warfare (NCW), where dispersed sensors, weapons, and decision makers across the battlespace enable synergistic, information-based effects.[9] NCW's primary advocate, Vice Adm Arthur Cebrowski, USN, retired, describes this capability as an RMA "unlike any seen since the Napoleonic Age, when France transformed warfare with the concept of *levee en masse*."[10] This near-perfect situational awareness implies fulfillment of the "endless quest for certainty" concerning command in war, and with such knowledge comes the expectation of decisive military intervention toward the achievement of one's political objectives: "what can be seen can be hit; what can be hit can be killed."[11]

Along with elevated situational awareness, an informational RMA supports also America's preference for wars of brief duration by leveraging technical speed. Information technology promises not only better knowledge but also faster decision making over a less capable adversary. This decision-cycle superiority translates into advantages in “tempo, initiative, and momentum” within the battlespace, promising “rapid mission accomplishment.”[12] Similarly, the technological component of an informational RMA within weaponry portends to be a panacea for another of America’s perennial political problems—casualty aversion. As the information content of the nation’s precision-guided munitions (PGM) continues to rise, so do “improvements in [their] range, lethality, and accuracy.”[13] This trend not only allows “the raw numbers of those weapons [to go] down” while maintaining the same capability, it also engenders the proliferation of unmanned platforms in the forms of standoff cruise missiles and combat aerial vehicles.[14] Thus, this combination of elevated speed in decision making, increased precision, and diminished risk to humans through information-enabled systems and weaponry meshes well with America’s general desire to minimize casualties and collateral damage during a short confrontation.

In reality, an informational RMA may harbor significant shortcomings. For example, perceived advantages stemming from the explosion and velocity of both data and data-sharing capability, as well as a potential compression of the command hierarchy, “can just as easily introduce confusion and become . . . liabilit[ies].”[15] Indeed, Clausewitzian friction retains a permanence in war regardless of how informational technologies may contribute to its marginal reduction. Informational RMA advocates minimize also “the existence of a thinking opponent who actively resists our plans, . . . seeks to achieve his own objectives,” and remains remarkably absent from their analyses.[16] Finally, informational RMA concepts of operation such as NCW are “optimized for force-on-force conflicts with conventional military opponents” rather than “today’s more ambiguous, asymmetric conflicts.”[17] Thus, although the power of information promises “many changes to the character of warfare, . . . we should not expect these changes to alter the nature of war” as defined by Clausewitz.[18] Nonetheless, ruminations of an informational RMA maintain traction throughout the national security apparatus, particularly due to its compatibility with former secretary of defense Donald Rumsfeld's vision of defense transformation.

Talk of an informational RMA reached its apex by the time Rumsfeld seized the reins of the Pentagon in 2001. Already the military services had struggled for years to grasp the ramifications of an informational RMA upon the efficacy of their national defense missions. Armed with his transformational image of an American military laced with technological and process improvements enabling force reductions, enhanced surgical-strike capabilities, and budget savings, Rumsfeld recognized immediately the compatibility of informational power to his agenda. Despite the armed forces' confusion surrounding the concept, IO emerged as one of six operational goals to facilitate transformation within the 2001 *Quadrennial Defense Review* (*QDR*).[19] Anointed by Rumsfeld as instrumental to defense transformation, IO regained its prior momentum as a "market . . . complete with catch buzzword[s] and Pentagon backing that each service scrambled to exploit."[20]

Information Operations' Detrimental Comprehensiveness

Interservice budgetary battles often spawn the proliferation of "new jargon" as service components attempt to influence the direction of a contested debate in their favor.[21] RAND studies observe that "adopting trendy language is a serious weapon in these wars," forcing interested parties to "talk the talk" while saddling affected mission areas with a surplus of overlapping and synonymous terms.[22] The concept of information within national security exhibits this affliction stemming from a fundamental lack of consensus concerning the meaning of the term itself. Depending upon one's perspective, information may serve as either the target or weapon in warfare. It may be considered a resource, a message, or a medium, and some even grant information status as a physical property of all matter akin to mass or energy.[23] These disparate points of view, along with "sloppy use of terminology at the most senior levels of the Department of Defense (DOD)," contribute to the "slicing, dicing, and boiling [of] the various manifestations of information" by those attempting to grasp or define its boundaries, yielding what Martin Libicki describes as "a lumpy stew."[24]

The hodgepodge that IO has become is exposed in stark relief within joint and service doctrine. Although attempting to elevate information as a legitimate component within the national defense arsenal, the DOD's *Joint Vision 2020* actually cripples the mission area with the dubious title of "key enabler" for defense transformation to achieve full-spectrum dominance.[25] This supporting, secondary status of IO translates into lukewarm institutional backing and an absence of focused leadership necessary to dictate and shape a common perspective. As a result, IO is described in Joint Publication 3-13, *Information Operations*, as "the integrated employment of electronic warfare (EW), computer network operations (CNO), PSYOP, military deception (MILDEC), and operations security (OPSEC), in concert with specified supporting and related capabilities, to influence, disrupt, corrupt or usurp adversarial human and automated decision making while protecting our own."[26] These actions generate information superiority, which in turn facilitates decision superiority, or the execution of "better decisions arrived at and implemented faster than an opponent can react, or in a non-combat situation, at a tempo that allows the force to shape the situation or react to changes and accomplish its mission."[27] Furthermore, joint doctrine attempts to incorporate all aspects of the information debate by acknowledging an information environment composed of physical, media, and cognitive dimensions.[28] To muddy the waters even more, many outside of the military categorize these core IO missions under the term *information warfare*, a subset of a broader notion of IO that integrates the other instruments of national power.[29] Regardless, such a broad panoply of characteristics and tasks attributed to IO hampers efforts to limit the breadth of the idea and, by extension, its application.

Within their respective service doctrines, both the US Army and Marine Corps build upon the five core IO mission areas outlined in joint doctrine. Recognizing that the joint definition of IO "brings together several previously separate functions . . . and related activities," the Army's Field Manual 3-13, *Information Operations: Doctrine, Tactics, Techniques, and Procedures*, distinguishes more explicitly between the offensive and defensive characteristics of IO.[30] This perspective requires not only extensive elaboration upon the seven desired effects from IO, such as denial or influence, but also the incorporation of defensive missions such as counterpropaganda and information assurance. Marine Corps Warfighting Publication 3-40.4, *Marine Air-Ground Task Force Information Operations*, adopts a similar stance but asserts that IO activities focus only upon the operational and tactical levels of war.[31]

The US Navy's guidance on IO in its Naval Warfare Publication (NWP) 3-13, *Navy Information Operations*, also attempts to build upon the joint construct but, instead, is the most nebulous of the four services. Although the five core mission areas are mentioned, NWP 3-13 allots a disproportionate amount of thought to organizational structure and command issues relating to IO. One helpful feature of the naval perspective is its aggregation of PA, CA, and PSYOP into an integrated concept focused upon influencing popular opinion; the document's emphasis upon IO's role in the strategic realm is distinctive.[32] Finally, the US Air Force's IO doctrine document, AFDD 2-5, *Information Operations*, endeavors to encompass all aspects of IO by not only expounding upon the five core mission areas but also addressing organization, supporting activities, and the information environment. Despite its broad nature, Air Force doctrine provides a useful construct with its categorization of IO into three capabilities—influence operations, EW operations, and network warfare operations.[33] Thus, the diverse and nonstandardized medley of America's IO capabilities throughout its services leaves the strategist unsure of how and where an information-based asset might mesh with an overarching military operation. However, each service's perspective provides a piece with which to decipher the puzzle of contriving a nonkinetic model for military operations.

Disaggregating Information Operations

Blinded by the flash of technology, anxious to harness information's promising power to alter the nature of warfare, and encumbered by interservice taxonomy tiffs, American defense professionals seem to have lost perspective of exactly what IO should emphasize. Infatuation with the speed of information has led to a compulsion to process and execute decisions faster than an opponent, a stance indicative of a kinetic mind-set. Furthermore, "attempting to use speed and knowledge to bring . . . conflict[s] to quick resolution" aligns nicely with the prevailing American vision of war.[34]

Although nonkinetic IO functions require the utilization of technology and physical infrastructure, they do not rely upon the speed of information and decision-making cycles to the degree some would contend. "IO is not something that can be done quickly or in a crisis mode"; it requires extensive lead times to ensure congruency with overall US security objectives as well as integration with other agencies.[35] Therefore, rather than dwelling upon the expansiveness and speed of information, which is theoretically limitless, "a back-to-basics approach that relates specific information-related tasks to broad . . . objectives" would restore focus to the military's blurred IO vision.[36] As such, the nonkinetic model of this study does not emphasize technology, nor is it indicative of the so-called informational RMA. Instead, a theoretically sound, nonkinetic model for military operations that reflects an accurate recognition of the unchanging nature of war and the importance of its moral component requires that "the various elements that are . . . lumped together as information warfare" be disaggregated.[37]

A back-to-basics approach to discerning the appropriate focus of IO must reexamine the fundamental purpose behind such activities. Fortunately, several categorizations of IO exist with which to accomplish this task. Jeffrey Cooper identifies three "crucial roots" to IO that characterize the historical strains of IO's core intent. The first and oldest strain emphasizes the molding of an enemy leader's strategic perceptions in the vein of Sun Tzu. Cooper's second root identifies signals intelligence and the exploitation of codes while his third historical strand is founded in electronic warfare and the manipulation of the electromagnetic spectrum.[38] These three roots equate roughly to a strategic-, operational-, and tactical-level focus for IO, although actions at any level may stimulate cascading effects into any or all of the others. The roots also grow from different components of an enemy's information system: its mind and the physical aspects of its network and transmission spectrum.

Michael Brown, an analyst for the secretary of defense's Office of Net Assessment, suggests a similar three-pronged functional approach to bound IO. First, perception management targets the "adversary's view of the world" and seeks to convince the enemy that any physical engagement comprises a hopeless endeavor. Next, information manipulation attempts to "sever the enemy's organizational nervous system" by disrupting communications networks, thereby hindering operational-level coordinated movements. Finally, information exploitation leverages sensor technology to elevate situational awareness of an enemy, increasing the probability of a successful kinetic strike.[39] Again, these functions encompass both mental and physical characteristics.

The USAF's contribution to deciphering the components of IO highlights three IO capabilities in AFDD 2-5. First, influence operations focus upon "the perceptions and behaviors of leaders, groups, or entire populations . . . to achieve desired effects across the cognitive domain." Second, network warfare operations emphasize the disruption of "any collection of systems [that] transmit information," whether human, digital, or analog. Finally, EW operations are conducted "across the electromagnetic spectrum" not only to coordinate and deconflict friendly use but also to attack and deny its use by the enemy.[40] These three IO capabilities follow the demarcations of both Cooper and Brown, and all three IO frameworks are found in tabular format at table 1.

Table 1. Information operations categorizations

Focus / Source	Psychological	Physical	
	Mind	Network	Spectrum
Cooper	Strategic	Operational	Tactical
Brown	Perception Management	Information Manipulation	Information Exploitation
AFDD 2-5	Influence Operations	Network Warfare Operations	EW Operations
Cox	Influence, Inform	Attack, Protect	Attack, Protect

Adapted from synthesis of listed sources.

Comprehensively, the Cooper, Brown, and AFDD 2-5 categorizations encompass the entirety of IO. However, if one accepts that the raison d'être of IO consists of "the influence of decision making of individuals, people in general, organizations, and . . . leaders," it stands to reason that those IO functions emphasizing the psychological domain carry the greatest importance.[41] Although actions targeting operational-level networks and the tactical-level electromagnetic spectrum unquestionably support an IO campaign, it is their second- and third-order aggregated effects into the mental, strategic-level realm that military strategists desire.[42] Furthermore, the network- and spectrum-centric elements of IO emphasize physical destruction and exploitation of adversary information processes and equipment for tactical or operational advantage, with little interest in changing how they or a neutral population think about a contentious issue. Finally, it is possible to generate mind-changing strategic effects via kinetic means; however, leadership strikes and decapitation strategies are not included in the nonkinetic model.[43] Such strategies belong within the kinetic model not only for their means but also for their focus upon destruction and compellence.

Therefore, to begin to separate a nonkinetic model from the mishmash of IO, only those functions designed to influence adversary or popular perception at the strategic level should be included as these activities relate directly to Clausewitz's moral element of warfare. Hence, the psychological column of table 1 represents the focus of the nonkinetic model, reproduced here as table 2. Extracting from the AFDD 2-5 complement of influ-

Table 2. Focus of the nonkinetic model

Source \ Focus	Psychological
	Mind
Cooper	Strategic
Brown	Perception Management
AFDD 2-5	Influence Operations
Cox	Influence, Inform

Adapted from synthesis of listed sources.

ence operations, the military capabilities of the nonkinetic model at this stage of analysis consist of PSYOP, MILDEC, OPSEC, counterintelligence (CI), counterpropaganda, and PA.

Another dimension that can deconstruct IO addresses its offensive and defensive aspects. US Army and Marine Corps doctrine contributes the most to this facet of a nonkinetic model stemming from their specific delineation of both offensive and defensive tasks within IO. Maj Joseph Cox, a 2006 graduate of the Army's School of Advanced Military Studies with recent service in OIF as an information officer, builds upon this perspective to offer an IO construct comprised of four functions: influence, inform, attack, and protect.[44] Cox not only recognizes the importance of both the enemy and the affected population in his influence function, he also carefully nuances this action from pure informing. His attack function emphasizes kinetic, tactical operations that may or may not support informing or influencing operations. Regardless of their intentions, the kinetic focus of these actions excludes them from a nonkinetic model. Although Cox allows for the fact that attack functions may take on nonkinetic forms, he mentions only computer network attack and MILDEC as examples. This perspective does not capture the entire breadth of offensive IO, which can consist of such tools as "PA, CA, [and] PSYOP" in both the prehostilities and active-engagement stages of conflict.[45]

The protect functions within Cox's framework possess two aspects, and both are largely nonkinetic in nature. The first seeks to defend friendly information capabilities, while the second endeavors to "protect the information environment friendly forces are trying to create."[46] Thus, the first aspect represents pure defense of friendly systems and perceptions, while the second view encompasses functions such as CI and counterpropaganda delivered via PA and PSYOP, which one can construe as offensive actions from a different perspective. Therefore, "it appears evident that both doctrinally and operationally, the distinction between offensive and defensive IO is becoming increasingly blurred."[47] In the effort to limit the scope of this study and despite the importance of protecting one's perceptions, networks, and the electromagnetic spectrum from manipulation and exploitation, the nonkinetic model will examine only the offensive aspects of IO. Included within this umbrella of activities are the technically defensive actions of PA and PSYOP delivered in reaction to adversary perception management. Thus, after identifying those IO actions aimed directly at the psychological element of warfare to achieve strategic effects, and separating or redefining those functions that are defensive in nature, only PSYOP, MILDEC, and PA remain to comprise the nonkinetic model.

Col William Darley, editor-in-chief of *Military Review* at the US Army's Combined Arms Center, as well as a career public affairs officer, categorizes PSYOP, MILDEC, and PA under the term *public information*, meaning that their "aim is to influence thinking processes and decisions through imparting ideas, words, patterns, or symbols; disseminat[ion occurs] through electronic or print medium, or word of mouth."[48] However, of the three functions, MILDEC stands out as most dissimilar. First, the intent of MILDEC is to mislead or deceive by depicting truthful information as false, or vice versa. This emphasis upon dishonesty and trickery cannot be supported in the long-term within a war of ideas. Second, MILDEC seeks "asymmetric advantage over an adversary through non-truth usually to support a specific tactical or operational kinetic action."[49] This lack of strategic focus, coupled with its principal use in support of kinetic action, serves to strike MILDEC from the repertoire of the nonkinetic model.

Finally, the US Navy's conception of "environment shaping" provides the final military piece to the nonkinetic model. "Environment shaping is the conscious action of molding the environment to prevent conflicts or placing US interests in a favorable position."[50] This process encompasses all actions taken to project a desired image and includes not only PSYOP and PA but also CA. Civil affairs' focus upon the establishment of civil relationships in friendly, hostile, or neutral areas is critical to erecting the legitimacy of an idea and in influencing popular opinion. Both PA and PSYOP may mutually support CA in this venture. Furthermore, the strengthening of host-nation capabilities to leverage indigenous resources that CA provides, to include self-policing, exerts long-term strategic effects to national and regional stability in a purely nonkinetic manner. For all of these reasons, CA activities, to include training, are addressed within this study's nonkinetic model whenever applicable. Thus, at this juncture of the analysis, the nonkinetic model consists primarily of PSYOP and PA, with the inclusion of CA when operationally feasible. A brief discussion delineating the differences between PSYOP and PA will solidify the intent and purpose of these activities.

Influence and Persuasion

The primary distinction between PSYOP and PA lies in the intended audience of each function. "Before and during any war, the government going to war must address three main target audiences: home, enemy (both military and civilians) and neutrals. For democracies, the most important audience is the home audience."[51] PA represents the nonkinetic capability focused upon the domestic audience, and its intention is to provide third-party access to known, knowledgeable officials for the dissemination of factual information. The power of public affairs, therefore, stems from the trust established between the messenger and the audience; any influence of popular opinion derives only from the audience's independent analysis of truthful, balanced information from a credible, legitimate source.[52] Cox's taxonomy describes this function as "informing," but he portrays the concept more passively than a proactive nonkinetic strategy may choose to employ it. More than "simply providing target audiences information on US activities, intentions, and operations," PA may seek proactively to allow high-visibility media access to spread specific facts designed to bolster domestic or international support.[53]

Unlike PA, PSYOP directs its message to foreign audiences with the sole purpose of shaping perception proactively to achieve an objective. These actions, although still truthful, focus more upon selling a message than disseminating factual pieces of information. Therefore, the source of the message is less important and may remain unidentified. Besides objective messages designed to elicit tactical or operational effects, such as methods of surrender or locations of humanitarian aid, PSYOP may also target thematic or subjective ideals intended to alter perceptions at the strategic level.[54] Cox describes these actions as "influencing," and he identifies correctly their intent to "change the behavior of a target audience." Furthermore,

Cox segments the foreign audience into two groups, suggesting PSYOP should focus upon enemy military forces during major combat operations while targeting the affected populace when embroiled in counterinsurgency operations.[55] Libicki offers an additional degree of fidelity and partitions PSYOP into four categories: counterwill (of the population), countercommander, counterforce, and counterculture. Although the first three are somewhat self-evident, counterculture encompasses the exportation of cultural products and ideals, such as recent US efforts to inject its political culture into foreign societies.[56] The nonkinetic model of this study embraces each of these views and will account for any method of PSYOP intending strategic effects.

Globalization has served to blur the lines of distinction between PA and PSYOP as messages intended for localized foreign listeners find their way back to domestic audiences. This phenomenon triggered the demise of the Office of Strategic Influence in February 2002, based on the admission that it may transmit disinformation. The DOD strategy of inserting pro-US articles into Iraqi newspapers in 2005 displayed similar tactics.[57] Both instances challenged the balanced, impartial nature of the press that the professional media strives to maintain and that PA requires to remain effective. Gen Richard Myers, chairman of the Joint Chiefs of Staff, attempted to separate PA from PSYOP in a 2004 memorandum, but even he included neutral, international audiences among the recipients of PA efforts.[58] Despite these challenges, PA and PSYOP remain viable, nonkinetic weapons within an arsenal of military capabilities. As with Thomas Paine's *Common Sense* that accelerated the momentum of the American Revolution by exerting "sudden and extensive" effects upon the American mind, dedicated emphasis upon truthful messages to persuade, coupled with balanced perspectives to influence, will ensure that any backfire associated with these nonkinetic actions in a war of ideas is minimized.[59]

Public Diplomacy

The final element of the nonkinetic model links the collection of PSYOP, PA, and CA to a political source. Since the kinetic model acts somewhat distinctly from politics, its nonkinetic counterpart should exhibit even greater continuity with political guidance. Therefore, a policy-grounded, nonkinetic method to express strategic intent—one that focuses foreign perceptions—would serve to lash up the nonkinetic model in its entirety. The addition of US public diplomacy to the elements of PA, PSYOP, and CA completes this study's nonkinetic model.

In general, public diplomacy combines elements of the diplomatic and informational instruments of national power in the effort to induce foreign publics to embrace the objectives of US foreign policy. Richard Kilroy synthesizes the definitions used at the US Department of State (DOS) and the erstwhile US Information Agency (USIA) to portray public diplomacy as "all those public information efforts officially endorsed by the US government aimed at informing and influencing foreign audiences, with the goal to shape their perceptions of America, Americans, and US foreign policy goals and objectives."[60] Therefore, public diplomacy differs from common conceptions of "soft power," which includes the allure of national values that are not overtly sanctioned by the government. Furthermore, the spectrum of soft power extends to the pole of passivity while public diplomacy endeavors remain generally proactive.[61] Thus, public diplomacy may be considered a subset of soft power, emphasizing persuasion and influence over co-option.

It is important to determine organizations responsible for US public diplomacy communications, and a brief historical examination here adds relevant context. As with most bureaucratic initiatives, public diplomacy enjoyed its greatest significance within US policy at the inception of its lead organization, the USIA, in 1953. In a major effort to streamline the United States' informational programs, the USIA expanded over the next eight years to incorporate the Voice of America (VOA) international broadcasting program; educational, cultural, and athletic exchanges; and processes to translate US books and journals into targeted foreign languages.[62] Pres. Ronald Reagan leveraged the USIA to transmit his strategic message of openness and resolve to areas of Soviet influence, and the eventual enlightenment and defection of previously closed societies can be attributed to some degree to these US public diplomacy activities.[63] Despite the apparent success of USIA efforts during the Cold War, the agency suffered during the subsequent reaping of the national peace dividend. Crippling budget cuts in the early 1990s portended the demise of the USIA, as many viewed its recent role as the counter-Soviet propaganda machine as obsolete.[64] Disbanded in 1998, the USIA's public diplomacy function transferred to the DOS under a new position—the undersecretary of state for public diplomacy and public affairs.[65]

During this time, the Clinton administration grappled with the vagaries of complex contingency operations in Rwanda, Haiti, and Bosnia, and in 1997 issued Presidential Decision Directive (PDD)-56, *Managing Complex Contingency Operations*, to rectify interagency coordination problems. Later, Clinton issued PDD-68, *International Public Information*, in April 1999, assigning the task of unifying national strategic communications to the new undersecretary of state for public diplomacy and public affairs.[66] Technically, this office continues to harbor the nation's public diplomacy role, although as the ultimate source of strategic communication, the sitting president dictates the strength of such a position.

The concept of "international public information" in the Clinton-era PDD-68 encompasses PA, international military information (a civilian pseudonym for PSYOP), and public diplomacy.[67] As such, the term represents well the components of this study's nonkinetic model on a strategic level. Furthermore, the military's role in support of US public diplomacy finds codification in a 2005 Defense Department directive that outlines the DOD's functions regarding stability, security, transition, and reconstruction operations. The directive defines *stability operations* as "a core U.S. military mission," encompassing the "support [of] indigenous persons or groups—political, religious, educational, and media—promoting freedom, the rule of law, and an entrepreneurial economy, who oppose extremism and the murder of civilians."[68] This institutional recognition of the military's pivotal role in stability operations removes any doubt concerning the military capability to contribute to US public diplomacy efforts. However, the key aspect that the public diplomacy element brings to the nonkinetic model is its function as the political anchor to which the subordinate military PA and PSYOP actions must cling. Without this source of overarching political guidance in the form of strategic communication from the executive branch, nonkinetic military efforts aimed at war's moral component possess little likelihood of finding their mark in a unified fashion.

Summary

This chapter arrives at a nonkinetic model in opposition to its kinetic counterpart. Contrary to many contemporary visions of IO that emphasize the technology of the information age, the nonkinetic model of this study sheds the multitude of electronic-based functions in IO to reveal its true raison d'être: the influence and persuasion of relevant audiences. America's fascination with the tactical and operational gains promised by the exponential increases in informational capabilities is understandable given their congruency with the nation's preferred methods of combat. However, this misguided emphasis obscures the strategic purpose behind IO that remains immutably independent of technology. By maintaining strict focus upon the psychological and strategic aspects of IO, the nonkinetic model meshes well with Clausewitz's moral element of war while simultaneously standing in contrast to the physical focus of the kinetic model.

Continued deconstruction of IO along the offensive and defensive dimension consolidates the relevant nonkinetic military capabilities into PA, PSYOP, and CA when feasible militarily. This process serves not only to simplify the terms of analysis but also to highlight the desired effects of influence and persuasion in opposition to the kinetic model's emphasis upon destruction and compellence. Finally, to grant nonkinetic military efforts an anchor in strategic policy and to further distinguish it from its kinetic antithesis, the nonkinetic model includes the function of public diplomacy.

Thus, the nonkinetic model of this study focuses upon the influence and persuasion of relevant audiences through the military capabilities of PA, PSYOP, and CA and the political guidance contained in public diplomacy efforts. These functions may employ information technologies, but they are not created by them. Furthermore, the nonkinetic model does not operate in isolation. Informational activities function within a dynamic environment that includes a rational, calculating opponent. Also, many of the psychological effects derived from the nonkinetic model's elements are enabled by kinetic means. Therefore, not only should nonkinetic actions anticipate adversary informational methods, they should also seek a synergistic relationship with supporting kinetic capabilities. In short, the nonkinetic model seeks to "substitute information for violence."[69]

Notes

1. Kurzweil, *Age of Spiritual Machines*, 20.
2. Steele, *Information Operations*, 28.
3. For explanations of this terminology, see Toffler and Toffler, *War and Anti-War*, 18–22.
4. Ibid., 4–5; and Gouré, "Impact of the Information Revolution," 230.
5. Libicki, *What Is Information Warfare?* 2–3.
6. Mazarr, Shaffer, and Ederington, *Military Technical Revolution*, 16.
7. Mets, *Long Search for a Surgical Strike*, 4.
8. Davis, "Information-Based Revolution," 83. See also Owens, "Technology, the RMA, and Future War," 63.
9. Moorman, "Future Role of Information Operations," 6.
10. Cebrowski, "Network-Centric Warfare," 29.
11. Van Creveld, *Command in War*, 264; and Allard, "Information Warfare," 238.
12. Hall, *Stray Voltage*, 59–60.
13. Krepinevich, "Keeping Pace," 24.
14. Allard, "Information Warfare," 238; and Lonsdale, *Nature of War*, 50.
15. Odom, *America's Military Revolution*, 48.
16. Hoffman, " 'New' Ideas," 29.
17. Ibid., 29.
18. Lonsdale, *Nature of War*, 232.
19. DOD, *Quadrennial Defense Review Report*, 30 September 2001, 30.
20. Munro, "How Private Is Your Data?" 14.
21. Buchan, "Information War?" 3.
22. Ibid., 3.
23. Cooper, "Another View of Information Warfare," 115; Steele, *Information Operations*, 2; and Arquilla and Ronfeldt, "Information, Power, and Grand Strategy," 135.
24. Bass, *Building Castles on Sand?* 2; and Libicki, *What Is Information Warfare?* 91.
25. *Joint Vision 2020*, 8.
26. Joint Publication (JP) 3-13, *Information Operations*, I-1.
27. *Joint Vision 2020*, 8.
28. JP 3-13, *Information Operations*, I-1–I-3.
29. Armistead, *Information Operations*, 19–20.
30. FM 3-13, *Information Operations*, v.

31. Marine Corps Warfighting Publication 3-40.4, *Marine Air-Ground Task Force Information Operations.*

32. Naval Warfare Publication 3-13, *Navy Information Operations.*

33. Air Force Doctrine Document (AFDD) 2-5, *Information Operations*, 5.

34. Metz, *Armed Conflict in the 21st Century*, 87.

35. Armistead, *Information Operations*, 113.

36. Buchan, "Information War," 3.

37. Ibid., 13.

38. Cooper, "Another View of Information Warfare," 112.

39. Brown, "Information Warfare."

40. AFDD 2-5, *Information Operations*, 5.

41. Hall, *Stray Voltage*, 98.

42. Ibid., 113–14.

43. Libicki, *What Is Information Warfare?* 9. Historically, such activities fell under the terms of *command and control warfare* or *kinetic persuade operations*. See also Darley, "Information Operations Theory," slide 80.

44. Cox, "Information Operations?" 3.

45. Armistead, *Information Operations*, 112.

46. Cox, "Information Operations?" 6.

47. Armistead, *Information Operations*, 113.

48. Darley, "Information Operations Theory," slides 100, 102.

49. Ibid., slide 104.

50. US Army War College, *Information Operations Primer*, 58.

51. Schleifer, "Democracies, Limited War," 41.

52. Darley, "Information Operations Theory," slides 115, 132.

53. Cox, "Information Operations?" 5.

54. Darley, "Information Operations Theory," slides 108, 132.

55. Cox, "Information Operations?" 5.

56. Libicki, *What Is Information Warfare?* 35–47.

57. Muravchik, "Hearts, Minds, and the War against Terror," 25.

58. Myers, "Policy on Public Affairs Relationship."

59. Ryan, *US Foreign Policy in World History*, 28.

60. Kilroy, "Public Diplomacy," 125.

61. Nye, *Soft Power*, 5–15.

62. US Information Agency, brochure.

63. Kilroy, "Public Diplomacy," 126.

64. Tolson, "War of Ideas," 32.

65. House, *Foreign Affairs Reform and Restructuring Act.*

66. Armistead, *Information Operations*, 125–31.

67. Ibid., 128.

68. DOD Directive 3000.05, *Military Support for Stability*, 2.

69. Steele, *Information Operations*, 3.

Chapter 5

Operation Enduring Freedom: The Models Applied

And when people are entering upon a war they do things the wrong way around. Action comes first, and it is only when they have already suffered that they begin to think.

—Thucydides

The collective shock inflicted upon America's psyche on 11 September 2001 impelled a series of national decisions that burdens today's military with the complications of irregular warfare in both Afghanistan and Iraq. After presenting the political objectives within the GWOT as the measures of military effectiveness, this chapter offers an analysis of the pertinent centers of gravity that the military instrument of power may affect based upon the identified threat. With these guidelines established, this chapter explores the ramifications of America's general military actions within OEF through the lenses of the kinetic and nonkinetic models. The chapter concludes with a brief overview of preliminary indications evident in OIF.

Political Objectives

The political objectives driving the military action within the GWOT are best derived from national strategy documents pertaining to the conflict. The national command structure has produced several unclassified publications of this nature, as indicated in figure 2. The *National Security Strategy* (*NSS*) deserves primacy in the determination of the political objectives of the GWOT simply from its position at the pinnacle of the hierarchy as well as its status as the expression of presidential guidance. Furthermore, the 2002 version of the document is most relevant to this study as its authority guided the earlier stages of OEF, while the 2006 edition builds upon the ideas espoused in its predecessor.[1]

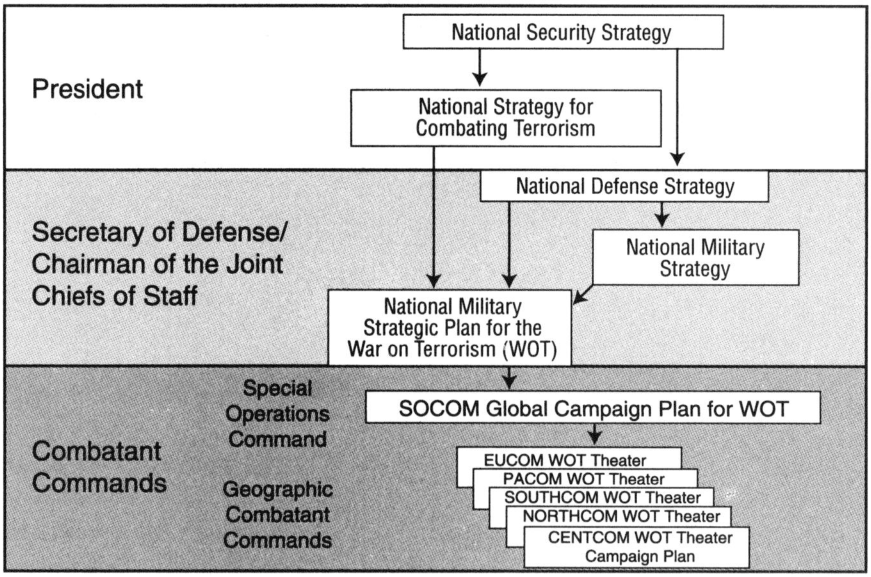

Figure 2. Strategic guidance hierarchy. (Reprinted from *National Military Strategic Plan for the War on Terrorism* [Washington, DC: Chairman of the Joint Chiefs of Staff, 1 February 2006], 9.)

Encouraged by the decline of totalitarian regimes, the president's 2002 *National Security Strategy* is "based on a distinctly American internationalism that reflects the union of our values and our national interests."[2] The term *internationalism* suggests a proactive policy of advancing American ideals abroad, while the confluence of values and interests spawns the stated national goals of "political and economic freedom, peaceful relations with other states, and respect for human dignity."[3] Similarly, America's *National Defense Strategy* (*NDS*) seeks to support the *NSS* by contributing to international conditions that include "the effective and responsible exercise of sovereignty, representative governance, peaceful resolution of regional disputes, and open and competitive markets."[4] Thus, the political objectives subsuming the GWOT can be expressed as the proliferation and protection of freedom, prosperity, and liberty.

From this superior guidance, several subordinate national strategy documents articulate varying visions of victory in the GWOT. The *National Military Strategy* (*NMS*) describes victory in the war on terrorism as its top priority to support the goals outlined within the *NSS* and *NDS*.[5] Furthermore, the *National Military Strategic Plan for the War on Terrorism* (*NMSP-WOT*) attempts to develop the notion of victory by illustrating the GWOT as "a war to preserve ordinary peoples' ability to live as they choose, and to protect the tolerance and moderation of free and open societies." This document describes the GWOT's end state as the reduction of terrorism from a severe, global threat to "an unorganized, localized, non-sponsored, and rare" phenomenon, restricted to the "criminal domain."[6] Likewise, the 2006 *QDR* asserts that victory occurs when terrorist networks lack the "ability or support to strike globally and catastrophically" while their regional capability is "outweighed by the capacity and resolve of local governments to defeat them."[7] This outcome requires not only "legitimate governments with the capacity to police themselves and to deny terrorists the sanctuary and the resources they need to survive" but also "support for the establishment of effective representative civil societies around the world, since the appeal of freedom is the best long-term counter to the ideology of the extremists."[8] Thus, the *NMSP-WOT* summarizes the two national strategic aims of the GWOT as the "defeat [of] violent extremism as a threat to our way of life as a free and open society" and the creation of "a global environment inhospitable to violent extremists and all who support them."[9] Despite their ambitious nature, these two officially declared political objectives serve as the standard by which military effectiveness in the GWOT should be gauged.

The Threat

In his 2002 *NSS*, Pres. George W. Bush establishes "terror" as the "enemy" in the GWOT, naming "premeditated, politically motivated violence perpetrated against innocents" as the most significant current threat to national political objectives. Such statements equate to waging war upon "fear," a psychological concept subject to relative interpretations. The document attempts also to personify the threat as "terrorists of global reach," and qualifies it as "not a single political regime or person, or religion, or ideology." This effort to clarify Bush's adversary of "terrorism" exacerbates the initial ambiguity further. To complicate the calculus an additional degree, the *NSS* "make[s] no distinction between terrorists and those who knowingly harbor or provide aid to them."[10] Likewise, the *NMSP-WOT* portrays a similar picture of the threat, painting the "transnational movement of extremist organizations, networks, and individuals—and their state and non-state supporters" as "the primary enemy" in the GWOT.[11] Sir Michael Howard expressed his incredulity with Bush's early guidance, reflecting that the "declaration of a 'war on terror' was generally seen abroad as a rhetorical device to alert the American people to the dangers facing them, rather than as a statement to be taken seriously or literally. . . . But further statements and actions . . . have made it clear that the President's words were intended to be taken literally."[12] Thus, national strategic documents attempt to identify positively the threat inspiring the GWOT; instead, a nebulous concept with which to guide military operations emerges.

The implications of a vague adversary within the GWOT are not all negative. Stephen Biddle suggests that "casting the net broadly makes it less likely that our war effort will inadvertently exclude important . . . threat[s] in an inherently murky domain." Furthermore, a broad threat definition may "create common cause with American allies facing terror threats of their own." Most importantly for America, however, Bush's expansive representation of the terrorist threat coupled with his characterization of the conflict as a "war" engenders "a moral clarity and normative power that helps marshal public support for the war effort."[13] This positive parameter likely represents the rationale behind Bush's decision to portray an ambiguous adversary within the GWOT. Thus, it seems presidential guidance in the early stages of the GWOT intended to blanket the conflict with a moral mantle designed to warm popular support.

Despite the ambiguity of the general terrorist threat within national-level GWOT guidance, President Bush does prioritize from the panoply of possible extremists by naming the perpetrators of the 9/11 attacks as the initial recipients of American retribution. During an address to a joint session of Congress on 20 September 2001, the president expressed that "our war on terror begins with al Qaeda, but it does not end there. It will not end until every terrorist group of global reach has been found, stopped, and defeated."[14] Thus, at least in the short-term, al-Qaeda and its associated movements characterize the manifestation of the extremist threat within the GWOT.

Despite the fact that "most Muslims are not fundamentalists, and most fundamentalists are not terrorists," the majority of contemporary terrorists are self-identified Muslims who exploit the tenets of Islam to garner additional adherents and support for their extremist cause.[15] Fundamentalists are not necessarily violent and comprise 10–15 percent of the Muslim population.[16] Similar to the conservative elements of Western Christianity, Islamic fundamentalists preach a return to the pure, authentic origins of their faith coupled with a corresponding rejection of modern theological distortions.

Although fundamentalists do not embody violent extremism, those who do typically emerge from these ranks. Known as Islamists, violent Islamic fundamentalists represent about 1 percent of Muslims who elect to elevate their agenda to the militant level.[17] Most closely represented by the Wahhabi sect of Sunni Islam, a faction founded within and backed by the Saudi monarchy, this minute segment of Islamic society signifies the "violent extremists" currently targeted within the GWOT, of which al-Qaeda is merely a subset.

The Islamist vision of a devout Muslim generates significant appeal among and within Islamic society for several cultural and ideological reasons. As a means to generate popular appeal, Islamists cite centuries past of Islamic regional domination, a Golden Age when their ideology and culture spread throughout the Arabian Peninsula, the Middle East, North Africa, and even Europe.[18] Experienced at a time recently removed from Mohammed himself, this era of prosperity and influence seems to coincide with the Muslim culture's greatest degree of congruency with the Prophet's teachings. Therefore, the Islamists assert, a proactive jihad to restore a fundamentalist transnational community should culminate in a similar period of glory based on Allah's favor.

This message found receptive ears in Osama bin Laden, and consequently al-Qaeda finds its primary impetus deep within the historical roots of Islam and the Islamist worldview. The al-Qaeda movement exploits this existential experience of the typical Muslim yearning for paradise lost. The movement promises divine recognition for the indiscriminate murder of civilians, and it spurs ideological fervor to repel infidel intrusion. Al-Qaeda's status as a violent extremist expression of Islamic fundamentalism is unquestioned; so also is the profound ideological basis for its very existence.[19]

Centers of Gravity

The *NMSP-WOT* provides a useful categorization of the nine "basic components" of an extremist network one may employ to unearth the probable centers of gravity of the GWOT.[20] While the relative importance of each component may shift over time within a conflict, its presence as a legitimate locus of military operations remains constant. As such, any or all of these areas of emphasis may be considered COGs, depending upon the specific threat, the characteristics of the adversary network, or other relevant contextual factors.

Finally, the nine elements described in the *NMSP-WOT* represent legitimate areas of operational emphasis for the US military. The elements exhibit characteristics of both the physical and psychological realms of warfare and may be affected simultaneously or separately by either kinetic or nonkinetic means. A brief synopsis of the nine components will underscore the military's capability to influence each, as well as reveal their relative importance within the GWOT.

First, *leadership* within extremist networks displays typically one of three functions: motivation, direction, or both. In the case of al-Qaeda, Osama bin Laden acts more as an icon of inspiration to the movement, while leaders of decentralized cells within the network dictate operational control. Military action may seek either to capture or kill extremist leaders or to undermine their credibility among a sympathetic populace harboring potential recruits. Both kinetic and nonkinetic means seem equally capable of affecting this network component.

Second, *safe havens*, whether physical or virtual, enable terrorist organizations to recuperate, train, organize, and plan. Physical sanctuary can be recognized and condoned by a host government or exist in uncontrolled regions external to governmental control. Military assets may target kinetically physical encampments to eliminate areas of refuge or coerce or remove sympathetic regimes to stimulate more effective host-governmental control. Nonkinetic military action may reveal the existence of terrorist sanctuaries to unsuspecting host nations or disrupt virtual safe havens electronically. Regarding the manipulation of this potential COG, the military's kinetic strength seems most capable.

Third, the exploitation of an extremist organization's *finances* may also cripple its ability to operate effectively. Whether funded by legitimate businesses or illicit criminal activity, military action may attack physically adversary sources of income or disrupt nonkinetically electronic monetary exchange or recipients of bribes. This focus of operation lends itself more to nonkinetic than kinetic action.

Fourth, *communications* comprise the reception, storage, manipulation, and dissemination of information. The military may influence this area of emphasis significantly through nonkinetic efforts. Whether through intelligence gathering, counterpropaganda, or electronic disruption of adversary communication, the military's nonkinetic arsenal stands particularly suited to target this component, as opposed to its kinetic counterpart that features attack of infrastructure.

Fifth, *movement* often proves critical to the success of an extremist operation and therefore represents a justifiable area of military emphasis. This component favors a kinetic response in that the very concept is grounded in the physical realm. Interdicting and preventing the movement of terrorists and their weaponry requires specific kinetic effects, although the creation of supporting documentation that enables surreptitious mobility may be vulnerable to nonkinetic military action.

Sixth, *intelligence* comprises both enemy exploitation of information to formulate and execute terrorist operations as well as the friendly ability to counter that intelligence. Military nonkinetic assets are well equipped to tackle this area of emphasis through deception and invasive electronic means.

Seventh, *weapons* are frequently a vital element of an extremist network susceptible to military pressure. Whether conventional arms, improvised explosive devices (IED), or WMDs characterize the arsenal of a particular terrorist organization, both kinetic and nonkinetic military operations contribute to halting the convergence of extremists and their preferred tools of destruction. Kinetic weaponry may aim to inhibit the manufacture or movement of terrorist armament while nonkinetic tools may affect adversely adversary training or impede the proliferation of key components through information exchange.

Eighth, the category of *personnel* represents the penultimate part of an extremist network that is vulnerable to either kinetic or nonkinetic assets. Both offer capabilities to reduce the membership of an extremist movement, but the military's nonkinetic efforts toward this potential COG promise superior success. Although kinetic strikes can only capture, kill, or isolate the individual branches of radical organizations, nonkinetic actions penetrate to their roots to starve the movement at its source: the legitimacy of their ideas.

Stemming from this phenomenon, the final and most critical component to radical networks is their *ideology*. This concept motivates every action of an organization's membership, from the decision to join to the commitment to kill innocents. This shared, comprehensive worldview inspires also sympathetic support from noncombatants and "sustains all other capabilities." This observation reveals the intransigence of ideology to a terrorist movement in contrast to the fungible nature of its other key components. Although the nature of the other eight components of an extremist network may be replaced or substituted with alternatives, an extremist's ideology cannot.

Accordingly, the *NMSP-WOT* elevates ideology as al-Qaeda's sole strategic COG within the GWOT, emphasizing the military's crucial, albeit supporting, role to restrict the widespread appeal of the Islamist message. In this area of operational emphasis, nonkinetic methods stressing the promotion of alternative ideas, the amplification of moderate voices, and the discrediting of adversary leadership are most appropriate. Although the US commitment to employ force may also stimulate pro-American ideological effects upon the moderates of a population, kinetic action must demonstrate considerable care and delicate application to avoid educing an unintended backlash of popular extremist support. Military contributions to the establishment of conditions that counter adversary ideology include population security; humanitarian and civil assistance; the meticulous, integrated implementation of culturally-aware operations; and information operations in support of national public diplomacy.[21]

Regarding the kinetic and nonkinetic analysis of the GWOT, any of these nine areas of military operational emphasis may be cited to explain the rationale and intended effect of a particular action. However, the character of OEF, coupled with the time limitations of this study, demand that certain components of the al-Qaeda movement receive elevated attention. Therefore, the analysis of kinetic and nonkinetic actions within OEF focus more upon the COGs of leadership and the safe havens embodied by sympathetic regimes, as they represent the administration's perceived paths to expedient political success in the GWOT. However, given that victory within the entirety of the GWOT "will come when the enemy's extremist ideologies are discredited in the eyes of their host populations and tacit supporters," judgments of the effects of military operations upon the COG of ideology receive the greatest priority in the OEF analysis.[22] Thus, within the GWOT, ideology represents the sole COG contributing to the achievement of political objectives over the long term, while exploitation of the subordinate COGs of leadership and safe havens can only muster short-term, temporary gains.

From these centers of gravity, the 2002 *NSS* outlines a general strategy for the prosecution of the GWOT. The nation's primary priority "will be first to disrupt and destroy terrorist organizations of global reach" by attacking their leadership; command, control, and communications; material support; and finances. Second, a subsidiary line of operation targets the war of ideas and includes "using the full influence of the United States . . . to make clear that all acts of terrorism are illegitimate [and] . . . viewed in the same light as slavery, piracy, or genocide." Third, a strategy to diminish the underlying conditions that spawn terrorism by "supporting moderate and modern government" through effective public diplomacy is also identified as contributory to the success of the ideological struggle.[23]

Thus, US national security strategy recognizes both the physical and moral components of the GWOT as well as potential missions for both kinetic and nonkinetic assets in its examination of the probable COGs. However, after analyzing these loci of operational leverage, civilian strategic guidance to the military within the 2002 *NSS* suggests the "disrupt and destroy" element of counterterrorism should take precedence over efforts to diminish its "underlying conditions." Whether such an operational emphasis represents an effective military strategy to "defeat violent extremism" and to "create a global environment inhospitable to violent extremists and all who support them" remains in question. An analysis of OEF from both a kinetic and nonkinetic perspective should shed some light upon this tenuous claim.

Operation Enduring Freedom

Presidential rhetoric during the prelude to combat operations in Afghanistan and language within national strategic documents issued in the ensuing aftermath display administration awareness of OEF's physical and psychological components. Furthermore, the magnitude of the ideological element to al-Qaeda's capability to inflict violence, coupled with its dearth of concentrated resources, suggests that the strategic concept surrounding OEF should have favored the moral end of the continuum of violence and featured nonkinetic lines of operation as the supported effort.

However, President Bush's address to the nation on the eve of hostilities reveals the administration's initial preference for a kinetic campaign to "disrupt the use of Afghanistan as a terrorist base of operations, and to attack the military capability of the Taliban regime."[24] This strategy emphasizes the physical-eradication portion of the "defeat violent extremism" objective and contributes also to the attainment of "a global environment inhospitable to violent extremists" by targeting a government facilitating terrorist activity. Thus, in its nascent stages, OEF portrays the US civilian leadership's decision to favor its kinetic arsenal against the physical elements of al-Qaeda's safe-haven COG provided by the Taliban regime. Leadership elements of both organizations were also targeted.

The Kinetic Model and Its Effects

Kinetic operations in the early stages of OEF display to some degree all of the characteristics of the kinetic model established in this study. The nature of the conflict reflected an American stance of compellence, and a destructive mind-set guided military operations. Some evidence of military activity disconnected from political objectives existed at the onset of hostilities, but this phenomenon became more prevalent around the summer of 2002. However, the US decision to employ its military force without overwhelming ground resources represents a departure from America's traditional kinetic construct, possibly derived from the promises of technological advantages coupled with indigenous forces. This compensation for a light US footprint on the ground, later known as the "Afghan model," seemed to garner effectively short-term physical gains, but left many underlying conditions of the conflict unresolved.

President Bush describes aptly the compellent nature of America's military operation in Afghanistan in a statement issued shortly after the commencement of hostilities: "I said to the Taliban, turn them over, destroy the camps. . . . I said, you've got time to do it. But they didn't listen. They didn't respond, and now they're paying a price. They are learning that anyone who strikes America will hear from our military, and they're not going to like what they hear. In choosing their enemy, the evildoers and those who harbor them have chosen their fate."[25]

Although some may consider this recollection of events as evidence of US negotiation with the Taliban, the statement reflects more the tone of an ultimatum with no room for compromise. Thus, 7 October 2001 marked the genesis of combat operations in Afghanistan with a military posture seeking either the complete eradication or the unconditional surrender of the Taliban regime and the elements of al-Qaeda under its protection.

Gen Tommy Franks's strategic concept centered upon three lines of operation emphasizing the destructive component of the kinetic model. The primary effort stressed "direct attack of the leadership of Al Qaeda and the Taliban" as well as "destr[uction of] the Taliban military."[26] The provision of humanitarian aid comprised the third, subsidiary focus of effort. Finally, plans to introduce operational maneuver with conventional forces were prepared should the requirement to infuse the operation with additional destructive power arise. However, early success of the Afghan model convinced many policy makers that significant follow-on US troops were not required. As events unfolded, OEF's kinetic operations assumed characteristics suggestive of a four-phase conflict.

The initial phase featured destructive air strikes against the Taliban regime and al-Qaeda leadership targets. Taliban "operation facilities such as radar, command-and-control centers and aircraft, as well as al Qaeda camps and headquarters" all received attention from US land- and carrier-based aircraft and Tomahawk cruise missiles.[27] The limited nature of Afghanistan's existing infrastructure coupled with its historical experience of decades of warfare allowed campaign planners to declare all major fixed targets destroyed by late October. However, mobile enemy forces of the Taliban and al-Qaeda, along with their leadership elements, "were . . . scattered, not eradicated."[28]

The second phase of OEF's kinetic campaign shifted operations from primarily fixed-target bombing to a concentrated, digitally-linked search-and-destroy effort against dispersed Taliban and al-Qaeda operatives. This "Afghan model" employed the indigenous anti-Taliban Northern Alliance in conjunction with special operations forces (SOF)-aided precision weaponry and proved effective in rooting out particularly stubborn pockets of resistance.[29] This methodology enabled Northern Alliance forces to march against the capital of Kabul five months ahead of predictions "after only twenty days of air strikes" and to capture it 24 hours later.[30] Early successes with the practice stimulated an influx of SOF teams into the country, with the initial allotment of three teams increased to 10 by mid-November, and again to 17 by 8 December 2001.[31] This expansion of the Afghan model precipitated a collapse of Taliban resistance around the country with Mazar-e-Sharif, Taloqan, and Jalalabad yielding to Northern Alliance forces in rapid succession, despite the increasing propensity of enemy forces to adapt to allied tactics.[32] Nonetheless, Pentagon officials estimated in mid-November that the Taliban controlled "less than one-third of the country, in contrast to 85 percent just a week before."[33] The final Taliban stronghold of Kandahar fell to allied troops in early December 2001, with Taliban leader Mullah Muhammad Omar and bin Laden fleeing into the hills.

The third phase of OEF transferred military focus from the Taliban strongholds and cities of Afghanistan to the cave network of Tora Bora along the Pakistani border. Over 1,000 al-Qaeda and Taliban operatives were tracked to the location, but only 200 US SOF existed in-country to assist in the search-destroy-and-capture campaign.[34] Again, the Afghan model template facilitated pinpoint targeting of occupied caverns, but the paucity of American troops necessitated the requirement for Northern Alliance forces to forestall al-Qaeda and Taliban infiltrations of Pakistan. Consequently, too much of the Pakistani border remained undefended, enabling the potential egress of enemy forces and high-level leaders such as bin Laden and Mullah Omar. Although the "most significant cave openings were destroyed and virtually all signs of live al Qaeda fighters disappeared," operations in Tora Bora concluded in mid-December with inconclusive results.[35]

The installation of the Afghani government under interim prime minister Hamid Karzai, an erstwhile tribal leader of the Northern Alliance, signaled the inception of OEF's fourth phase that persists to this day. Announced by Secretary of Defense Rumsfeld on 21 December 2001, the new phase sought to protect the fragile Karzai government from Taliban and al-Qaeda resistance through continued, kinetic "search-and-destroy" operations "to capture or kill" any remaining enemy fighters.[36] Operation Anaconda represents perhaps the most well-known military activity of this phase and commenced in southeastern Afghanistan on 1 March 2002. Anaconda focused upon "hundreds of suspected Taliban and al Qaeda holdouts in the Khost and Patika provinces" and relied upon an "increased number of conventional forces" in addition to SOF and Northern Alliance elements.[37] Despite its exclusion during planning and initial execution, airpower "was summoned at the eleventh hour" when failure seemed imminent and following unexpected levels of Taliban and al-Qaeda resistance.[38] After several days of bombing, the operation was declared a success after adversary forces dispersed; however, as at Tora Bora, "hundreds of . . . fugitives escaped into Pakistan," stemming from the inability of proxy forces to seal off vital avenues of escape.[39] More importantly, bin Laden and Mullah Omar remained unaccounted for.

The fourth phase of OEF marked also the increased divergence of US military operations from expressed political objectives, as the United States' historical employment of the kinetic model predicts. July 2002 found evidence of a rising insurgency in Afghanistan, with guerrilla attacks upon US military and interim Afghani government targets growing in frequency. This alarming trend instigated a US response toward elevated employment of conventional forces and tactics, culminating in August's Operation Mountain Sweep. This action witnessed hundreds of US 82nd Airborne infantry comb thoroughly areas along the Pakistani border believed to harbor the insurgent base. Although these conventional troops succeeded in securing arms caches and "more than a dozen suspected Taliban affiliates," their invasive, culturally-ignorant methods reportedly "terrorized" the affected villagers and damaged the "trust that SOF had worked for months to gain."[40] Instances like these did little to engender support for the fledgling Karzai government, thereby bringing the operation's contributions to the political objective of creating "a global environment inhospitable to violent extremists and all who support them" into question. Furthermore, it was only after the negative effects of kinetic operations such as Mountain Sweep became apparent that operational emphasis began to shift toward the reconstruction and long-term stabilization efforts necessary to sustain a positive platform in an ideological struggle.

The highlighting of military divergence from long-term political goals at this point should not suggest that operations prior to Mountain Sweep exhibited complete congruency with political aims. Within a week of the initiation of bombing in October 2001, a Red Cross warehouse absorbed an American precision air strike. Although collateral damage will remain inherent in warfare for time eternal, the absence of agency coordination distinguishes this event from other instances of errant attacks. The warehouse was not struck inadvertently; rather, the structure was targeted intentionally as a potential storage site of enemy supplies despite the longevity of its use and the existence of clear markings by the Red Cross.[41] Although this event did not impact significantly Red Cross efforts, the fact remains that simple deconfliction of kinetic actions from long-standing humanitarian efforts in advance of hostilities did not occur. Unfortunately, "what had been an easy working relationship before the war between U.N. [United Nations] agencies and NGOs [nongovernmental organizations] declined in the absence of an integrated mechanism for coordinating U.S. military objectives and Afghan or international civil objectives."[42] This breakdown of communication reveals the slight attention devoted to moral considerations that contribute to the foundation of legitimacy in a war of ideas, as well as the primacy of the kinetic campaign in the early stages of OEF.

In sum, the kinetic campaign of OEF yielded undisputed, tangible gains. US and allied territorial control expanded continually throughout Afghanistan until all major population centers succumbed to their forces, and in just weeks, the Taliban ceased to function as a government. Kinetic efforts destroyed 11 terrorist training camps and 39 Taliban command and control sites.[43] Taliban personnel losses ranged from 8,000 to 12,000, with another 7,000 captured. Egress routes into Pakistan were largely blocked, and prior to OIF one could claim confidently that "al Qaeda is weaker without its Afghan sanctuary."[44] It seemed American and Northern Alliance kinetic efforts had exploited masterfully al-Qaeda's safe-haven COG of the Taliban regime. Despite these considerable physical gains, many underlying moral conditions remained unresolved or were exacerbated by kinetic operations that allowed resistance to persist.

The kinetic approach's inability to capture or kill the principal leaders of both al-Qaeda and the Taliban regime represents its primary shortfall. Derived from US impatience with the relatively slow deployment rates of conventional Army units into the austere Afghan environment, or the political decision to limit US casualties during initial ground operations, US reliance upon indigenous ground support netted short-term physical gains at the expense of long-term political consequences. Although US lives were undoubtedly saved and initial operations expedited by capitalizing upon Northern Alliance availability and expertise, its questionable commitment to US political objectives may have contributed to its lackluster border-control efforts at Tora Bora and during Operation Anaconda. The capability of key al-Qaeda and Taliban leaders to inspire and cultivate the ideological fervor fueling Islamism comprises a critical source of enemy power, and departure from the kinetic model's traditional emphasis upon overwhelming force permitted that strength to persist.

The many breaches of popular security now plaguing Afghanistan embody a second political ramification of leveraging local ground support. The lack of population security from armed harassment hinders current reconstruction efforts and impedes political normalization. Paradoxically, many of the offending belligerents are the same tribal "warlords" that the United States employed during larger-scale hostilities.[45] Thus, the same factions that initially banded together to oust the Taliban and install the Karzai administration now serve to delegitimize the Afghani government by preventing the spread of its influence via reconstructive efforts. Again, this phenomenon is at least partly attributable to the kinetic-based decision to pursue short-term physical gains at the expense of long-term political stability.

The kinetic approach's willingness to accept collateral damage in certain instances, while failing to mitigate its effects in others, comprises another strategic deficiency in the ideological struggle of the GWOT. Pursuant to their kinetic strategy concerning the leadership of both the Taliban and al-Qaeda, US military planners elected to target "residences and residential areas" whenever intelligence suggested the presence of a high-level operative.[46] Perhaps lulled by the promise of precision weaponry, this tactic of urban targeting provided the Taliban with a multitude of opportunities to highlight the inadvertent slaying of innocents. As early as 10 days into the operation, the tenuous allied coalition exhibited cracks from one-sided collateral damage reports when Prince Naif, the Saudi interior minister, commented that his kingdom "opposed terrorism, but did not approve of the US response [that] is killing innocent people."[47] The United States exerted little effort to lessen the moral effect of Taliban claims of civilian deaths at this stage. US officials such as Secretary of Defense Rumsfeld admitted that, regrettably, these acts comprise the price of combating terrorism, or they simply refuted Taliban claims as outright fabrications.

Initial field reports suggest that the Afghanistan bombing campaign directly claimed more than 1,000 civilian lives. Although some tout this figure as evidence of the "relatively modest harm to innocents" in the early stages of OEF, the total represents one civilian death for every 12 bombs or missiles expended.[48] By comparison, the 1999 Balkans campaign witnessed one civilian death for every 46 bombs dropped.[49] Unfortunately for the United States, expressions of the superiority of precision weaponry bring with them heightened expectations of truly surgical strikes, and the United States proudly quoted the metric that 70 percent of its expended munitions utilized precision technology.[50] However, the economical decision to favor the global positioning system–guided Joint Direct Attack Munition (JDAM) system over other more expensive, but more accurate, laser-guided weapons signifies American acceptance of greater risk of collateral damage and long-term loss of popular support in order to prosecute an aggressive, air-based kinetic campaign against adversary leadership.

Although the US handling of collateral damage claims improved over time, the persistence of the kinetic campaign beyond its initial utility contributed to the magnification of adverse effects surrounding unintentional civilian deaths. July 2002 witnessed an AC-130 gunship attack of an Afghani wedding celebration in response to traditional celebratory gunfire. Locals claimed 48 dead and 117 wounded, and the negative publicity stemming from US cultural ignorance impelled US military officials to advocate a ground-based approach featuring "more civil affairs and security teams to provide aid and regain American rapport in the [Uruzgan] area."[51] This construct formed the basis of the provincial reconstruction team (PRT) concept and marked the United States' stubborn recognition of the growing need to rectify more actively and prevent kinetic-based collateral damage and to refocus its efforts upon the psychological element of warfare.

Unfortunately, kinetic actions producing significant civilian casualties continue, contributing to a "sense of insecurity and fear in the country" and undercutting the credibility of the Karzai administration.[52] As recently as May 2007, SOF-directed air strikes claimed responsibility for 130 Taliban deaths near Herat that local officials, including a member of the Afghani parliament, insist were civilians unaffiliated with the former regime.[53] The same month found the United States issuing payments of restitution to 69 Afghani families victimized by human losses during a Marine special forces' indiscriminate retaliation of an attack upon their convoy. The commander of the offending unit, Col John Nicholson, recognized that "the people are the center of gravity here, so, first and foremost in all that we do, we seek to do no harm to the people."[54] This observation asserts indirectly that popular support of the US-backed Karzai government remains critical for US credibility in the greater war of ideas. Thus, the US practice of operating under kinetic "rules of engagement that were more appropriate to the intensive days of the war [seem to add] to the acute embarrassment of the Karzai government and at serious cost to its political viability" during OEF's later stages.[55]

The most damning statistics pertaining to the kinetic-centric US strategy in Afghanistan, however, relate to the glacial rate of infrastructure improvements that likely instigates the increasing incidences of violence born of popular frustration. Despite concrete advances in political and educational reform, a September 2006 World Bank report relates "only 13% of Afghans have access to safe water, 12% to adequate sanitation, and just 6% to electricity. An estimated 7 million people remain vulnerable to hunger despite increases in agricultural production. . . . More than 70% of schools need repairs, as do most of the country's primary roads. Life expectancy is 44 years (compared to 59 years for low-income countries worldwide)."[56] Ground travel from Kabul to Kandahar that once took three hours now consumes 14.[57] After almost five years of US and Karzai claims of credibility and capability, such statistics do little to sway popular opinion in an ideologically-driven conflict.

Unsurprisingly, indications of elevated frequency and degrees of violence now exist in Afghanistan. After a lull in the overall level of violence during the spring/summer 2002 time frame, an increase of insurgent guerrilla-like tactics is evident, particularly those employing IEDs and suicide attacks. "Suicide attacks, formerly rare in Afghanistan, have risen from five in 2002–2004 to 17 in 2005 and 14 in 2006."[58] These incidents mirror those in Iraq, as does the May 2005 attack of a mosque in Kandahar yielding a significant death toll. Likewise, in July 2005, "captured Afghani police were beheaded by insurgents."[59] The nature of these actions seems to have migrated from Iraq and percolates in an environment populated by individuals dissatisfied with the services provided by a so-called reformist government.

Furthermore, the prolonged US emphasis upon a kinetic, physical strategy that alienates locals to an increasing degree inhibits the impetus of reconstructive efforts and limits the regional influence of the Karzai administration. In 2004, the situation deteriorated to the extent that "half of the country's 32 provinces [were] no-go areas for aid workers"; however, a belated push toward a nonkinetic approach centered upon reconstruction is slowly shifting the momentum.[60]

Thus, from the kinetic perspective, "the major goals of [OEF], to overthrow the Taliban, and to reduce the influence of al Qaeda, were achieved."[61] The al-Qaeda safe-haven COG personified by the Taliban regime ceased to function as a governing entity, and both organizations' leadership and communications structures were at least disrupted. Kinetic actions seemed to support the established political objectives of defeating violent extremism and creating an environment hostile to its persistence. The strategy achieved its aims in the short term. However, enough remnants of Taliban and al-Qaeda forces remained in the country to affect adversely the long-term stability of the region, and their key leadership eludes capture.

The United States missed an opportunity to shift its operational focus upon the strengthening of the Karzai administration shortly after its installation. The summer of 2002 witnessed the continuation of the US kinetic approach despite evidence of an embryonic insurgency. Rather than swing explicitly to a nonkinetic priority of effort to strengthen the legitimacy and credibility of the Karzai government, kinetic-centric operations continued to the detriment of popular support. Although US attention did move toward Iraq during this time frame, America displayed reluctance at this critical juncture to pursue aggressively the conditions for long-term success in Afghanistan; the United States allowed conditions to deteriorate by failing to build ideological credibility through tangible reconstruction.

Recognition of this failure occurred at a later date, but lost Afghani trust requires that much more time and effort to restore. "A disturbing trend is the belief among some in OEF that the Coalition is barely breaking even in the information war."[62]

The Nonkinetic Model and Its Effects

Despite Bush administration recognition of the ideological element within the GWOT, diminished emphasis on nonkinetic resources is starkly apparent throughout OEF. From ambiguous yet inflammatory strategic communications that lacked a supporting bureaucratic structure to refine the American message, to military PSYOP and CA efforts that suffered from little to no direction from higher command echelons, US nonkinetic endeavors contributed little toward the exploitation of the ideological COG of the GWOT. Realization of this fact garnered greater traction, however, following the adverse popular opinion surrounding US conventional operations in the summer of 2002. The emergence of PRTs in early 2003 represents a positive, albeit belated, strategic move toward leveraging the ideological COG with nonkinetic effects.

In the public diplomacy realm, President Bush displayed his early misunderstanding of the importance of the ideological COG in the GWOT with his characterization of the impending conflict as a "crusade." Informing America that "this crusade, this war on terrorism, is going to take awhile," Bush provoked instantly the ire of Muslims everywhere sensitive to the historic connotations of the religiously-charged term.[63] Recognizing that US strategic communication may have inadvertently alienated moderate Muslims central to muting Islamism as well as given terrorist organizations of disparate ideological backgrounds a "powerful reason to cooperate," the Bush administration scrambled to clarify its message.[64]

President Bush appointed Madison Avenue advertising executive Charlotte Beers to the post of undersecretary of state for public diplomacy and public affairs in an effort to "rebrand" the US image abroad in general and its GWOT message in particular.[65] Beers's "major product was a shiny and colorful 25-page pamphlet, *The Network of Terrorism* . . . featuring vivid photographs of the September 11 destruction." Prominent quotations from the Council on American-Islamic Relations (CAIR) appear throughout the publication repudiating the 9/11 attacks; however, it was discovered later that the leader of CAIR was an avid proponent of Hamas and Hezbollah.[66] In this instance, "the messenger may be more important than the message," and efforts to provide a Muslim voice to a US message fell flat.[67] Beers also produced *Muslim Life in America*, a publication attempting to illustrate the compatibility of US democracy with its Muslim citizens.[68]

Although this message reiterated the United States' positive views toward mainstream Islam, the perspective emphasized all that is right with America rather than what is wrong with the extremist ideology. Furthermore, throughout OEF the US public diplomacy message continued to aim "at influencing morale and support for the war in the United States" instead of focusing upon energizing the moderate Muslim populace.[69] Finally, the Bush administration focused upon "clamping down" negative press emanating from external sources such as the Qatar-based Al Jazeera television network, rather than "fighting back" with its own communications efforts.[70] Thus, early US attempts to articulate a strategic message stressed the positives of America's stance, focused upon mostly domestic and non-Muslim international audiences, and attempted to stifle two-sided commentary rather than engage in an informed debate.

Along with America's misguided message, it became evident quickly that the nation's public diplomacy apparatus suffered from the absence of an effective bureaucratic structure designed to transmit a unified voice. Attempts to invigorate the US public diplomacy arm shortly after 9/11 revealed that "the institutional expertise that [once] skillfully managed information programs for foreign audiences . . . no longer exists."[71] The merging of the USIA into the State Department in 1999 "had damaged overall US public diplomacy efforts by cutting valuable resources . . . and undervaluing the mission of public diplomacy in supporting US national security objectives."[72] Victimized by the DOS's subservience to the tyranny of the urgent, US public diplomacy lost its long-term focus, becoming disintegrated from US foreign policy.[73]

A key message-coordinating entity, the Office of Strategic Communication represented one casualty of the supposed cost-saving consolidation that DOS decision makers deemed worthwhile.[74] President Bush attempted to rectify this bureaucratic lacuna by establishing the Strategic Communications Policy Coordinating Committee within the White House staff in 2002. However, the group disbanded a year later, unable to fulfill its mandate to "draft a national communications strategy."[75] The administration erected next the Office of Global Communications in January 2003 to handle message unification, but as of May 2006, no top-level public diplomacy guidance has been issued to lateral and subordinate agencies.[76] During the period from October 2001 to March 2003, the paucity of DOS public-diplomacy guidance to its field offices propelled some locations to formulate intracountry communication plans based upon "newspaper articles and guesswork."[77]

A November 2001 Council on Foreign Relations (CFR) study confirmed the dilapidated state of America's public-diplomacy program, finding that "public diplomacy is a low bureaucratic priority as reflected by the relatively low-level officials traditionally assigned to it and the meager resources normally allocated to it."[78] Funding for US public diplomacy plummeted since the 1999 USIA merger into the DOS, with its $2 billion annual budget slipping to only $500 million by 2004.[79] This period witnessed the slow weakening of the Voice of America broadcasting service in predominately Muslim regions, as transmissions dwindled to only seven hours per day in a single Arabic dialect that reached an estimated 2 percent of the Arab world.[80] Although public-diplomacy funding has risen about 21 percent over the past two years, "staff numbers have stayed largely the same," thereby limiting the institutional reach of the additional funds.[81]

Besides consolidating the USIA into the DOS, the 1998 Foreign Affairs Reform and Restructuring Act also "spun off foreign broadcasting as an independent entity."[82] The Broadcasting Board of Governors (BBG) assumed control of the entirety of the US government's international broadcasting effort, including the VOA. This politically-charged body, comprised of four political appointees from each party as well as the secretary of state, frequently struggled to assert its bureaucratic authority, instituting policy changes "just to show that it is in charge."[83] Inspired by the 1948 Smith-Mundt Act that simultaneously authorized the "dissemination abroad of information about the US, its people, and its policies" while prohibiting those same messages to domestic audiences, the BBG viewed its mission as distinct from political interference.[84]

This organizational construct contributed to bureaucratic turf battles that hindered the unified expression of America's message. For instance, shortly after 9/11, the new BBG appointed a Bush administration candidate to direct the VOA; some BBG members "then allegedly undercut his decisions, resulting in a resignation" and distracting scandal at a critical period.[85] Furthermore, VOA aired days later an interview with Taliban leader Mullah Omar laced with "venomous attacks against the US."[86] Not only did this event undermine US informational efforts, it also occurred in opposition to DOS guidelines and resulted in the firing of the VOA director.

There exists no consensus concerning the efficacy of BBG programs instituted since 9/11. On the positive side, the BBG "established the Middle Eastern Radio Network (MERN) and Radio Sawa broadcasting 24 hours a day in Arabic on AM, FM, shortwave, digital satellite, and the Internet . . . just six months after the September 11 tragedy."[87] Furthermore, "Radio Farda . . . began transmitting into Iran . . . in December 2002" with similar capabilities, and the Afghanistan Radio Network was launched in February 2003.[88] These media channels are designed to transmit truthful reports of American policy to the Muslim world. As an example, Secretary of State Colin Powell's speech to the UN in February 2003 laying out the US case against Iraq was broadcast on these outlets "as it was delivered."

BBG "informal survey data" show that Radio Sawa is the "most popular station in many Arab capitals" and that the service reaches 51 percent of its target audience.[89] Based on these favorable reports, Congress authorized the expenditure of $62 million to establish Al-Hurra, an Arabic television network intended to counter the coverage of Al Jazeera.[90] Despite the undisputed increase in transmission of US opinions throughout the Arab community, these programs still suffer from the lack of superior guidance to ensure consistent messages to overseas audiences.

DOS studies suggest that the actual effects of BBG programs on public attitudes have yet to be determined. The state inspector general concluded in 2004 that the BBG "focused more on audience size and composition than on potential impact on attitudes in the region."[91] Others contend that "the $62 million spent on [Al-Hurra] would have been better used purchasing 'quality American content' for indigenous Arab satellite networks," and the executive editor of the *Daily Star* in Beirut calls Al-Hurra and Radio Sawa "an entertaining, expensive, and irrelevant hoax."[92] Finally, other detractors of BBG efforts cite their foundation upon a Cold War model designed for European audiences as evidence of the programs' "irrelevan[cy] to the Arab world."[93] Until adequate methods of measuring the impact of BBG-derived US public diplomacy programs are created, the true worth of these efforts remains in question. Thus, as the United States commenced military operations in Afghanistan, American leaders became painfully aware of the ramshackle and disintegrated state of its public diplomacy architecture.

Not until 8 April 2006 did President Bush establish a Policy Coordination Committee on Public Diplomacy and Public Affairs to ensure "all agencies work together to disseminate the President's themes and message; all public diplomacy and strategic communications resources, programs, and activities are effectively coordinated to support those messages; and every agency gives public diplomacy and strategic communications the same level of priority that the President does."[94] Such an organizational move nearly five years after 9/11 illustrates the incredible magnitude of America's bureaucratic inertia regarding public-diplomacy reform.

Consequently, America's ability to affect positively the pivotal strategic COG of ideology in the GWOT remains restricted, sporadic, and disintegrated. As Lisa Curtis, senior research fellow at the Heritage Foundation asserts, "We have our work cut out for us . . . opinions of America have declined markedly—to all time lows in some countries—over the last few years."[95] This trend exists despite institutional recognition of the importance of the problem and legitimate efforts to alter Muslim perception of America. It seems the lack of emphasis upon the nonkinetic aspect of OEF and the GWOT has allowed US public-diplomacy programs to flounder under a misguided message centered on

America rather than Islamism while remaining shackled to a haphazard bureaucracy incapable of sufficient reform.

Despite the lack of a unified voice guiding US public diplomacy from the executive branch, the DOD attempted its own programs to rectify the strategic deficiency. Acknowledging the fact that the DOD "plays a supporting role in public diplomacy," the department's Defense Science Board commissioned a task force in October 2001 on "managed information dissemination" that recommended the creation of a National Security Council policy-coordinating committee after discovering the "understaffed, underfunded, poorly coordinated, and insufficiently integrated" state of US public diplomacy programs.[96] The administration adopted the recommendation, but no tangible guidance emerged from the short-lived committee.

Frustrated by the absence of strategic informational guidance, the Pentagon developed plans "to provide news items, possibly even false ones, to foreign media organizations" under the direction of a DOD Office of Strategic Influence (OSI).[97] This admission, in the form of a press leak, ruffled the feathers of citizens familiar with the 1948 Smith-Mundt Act; they expressed concern that the military roles of public affairs (influence) and PSYOP (persuasion) would become indistinguishable. Caving to political pressure, Secretary Rumsfeld jettisoned the OSI endeavor before its genesis, forcing the DOD to seek alternative methods to focus its informational programs.

Turning to independent contractors, the DOD chartered the Rendon Group (TRG) to assist the department in tailoring its media products. Soon after 9/11, this strategic communications consulting firm conducted media analysis of Arab news services and other Middle Eastern broadcasts to provide recommendations to the DOD "on how to counter anti-American themes and messages" in Arab media outlets.[98] Furthermore, TRG supports also the DOD's classified IO campaigns directed by US Strategic Command at Offutt Air Force Base, Nebraska. Again, although the effects of these efforts are difficult to measure, moves toward the privatization of DOD informational campaigns enable the military to pursue programs with a lesser degree of public scrutiny. The United States' lack of informational awareness and consistency in the public diplomacy arena translated also into the military public affairs role.

The US decision to conduct its public affairs briefings from the Pentagon rather than a deployed location exhibited slight regard for adversary efforts to control the information war. The practice tended to delay US reports from the field as Pentagon officials digested and repackaged the previous days' operations.[99] This placed the primary messenger of combat information to America and some parts of the world in a reactive posture and "allowed the Taliban to quickly seize the . . . initiative in collateral damage reporting."[100]

Swayed more by the timeliness of Taliban reports rather than by their accuracy, US media continually pressured Pentagon officials to refute or verify Taliban claims. The Pentagon's evasive and often undiplomatic responses did little to alter perceptions of US bombing effects, and world opinion of US strategy in OEF plunged. By the end of October 2001, 57–70 percent of Turkey's population opposed direct involvement in OEF despite its standing as the only Muslim nation of the North Atlantic Treaty Organization (NATO). Furthermore, half of the Kuwaiti parliament called for an end to the bombing, and 50 percent of the Middle Eastern respondents in a Gallup poll viewed US military action in Afghanistan as "morally unjustifiable."[101] These trends are not surprising given the reactionary informational environment the United States allowed to persist during OEF.

The Rendon Group identified this phenomenon and instituted programs to rectify the reactionary reporting. TRG worked with British and Pakistani allies to establish 24-hour information centers to function on a continuous news cycle and attempt to lead war-related coverage. This practice proved effective, and the playing field of collateral-damage reporting began to level.[102] Unfortunately, the DOD adapted itself to the hostile information environment with little support from national public diplomacy programs. The aforementioned 2001 CFR report observed "the federal bureaucracy is not configured to handle the demands of a major public diplomacy campaign, . . . and [to be] successful, [a] public diplomacy campaign will . . . need to be nimble enough to take advantage when a situation presents itself, and fast enough to respond to negative charges before they can take hold in the popular imagination."[103] The military now understood this fact first-hand, but not until the winter of 2005 did the State Department erect a "rapid response unit that follows newscasts around the world and offers talking points on breaking international news to rebut negative media stories about the US."[104]

This US informational embarrassment exists astride an al-Qaeda ideological campaign exuding outstanding focus and direction. From bin Laden's initial fatwas delineating al-Qaeda's specific religious and political aims, to the high-quality production of its video and Internet propaganda, al-Qaeda treats the informational war as the primary battlefield.[105] Transmitting a refined message comprised of a "clever mix of history, psychology, and paranoia," al-Qaeda's media program capitalizes upon the accessible audiences of Al Jazeera, and its use of American spokesmen such as Adam Yehiye Gadahn reflects a concerted desire to bolster its Western recruitment efforts.[106] Finally, Ayman al-Zawahiri's admonition to Abu Mussab al-Zarqawi during the latter's escalation of sectarian violence in Iraq exhibits impressive awareness of the requirement to preserve the unified nature of al-Qaeda's anti-Western message.[107] From the public diplomacy perspective, al-Qaeda's skillful ability to transmit and refine a consistent, receptive message dwarfs US attempts to do the same.

Military nonkinetic actions in the field during OEF contributed primarily to kinetic operations, although indirect and direct contributions toward the COG of ideology also transpired. The principal focus of PSYOP activity at the onset of OEF centered upon supporting the overall kinetic emphasis upon the COG of the Taliban regime. Leaflet drops occurred throughout hostile areas, attempting to influence enemy combatants to surrender. These products emphasized the futility of armed resistance in the face of overwhelming coalition firepower and did not target Taliban or al-Qaeda motivations for fighting. As such, enemy surrenders were few. EC-130E/J Compass Call broadcasts offered incentives for local Afghans to provide information leading to the capture or killing of key Taliban or al-Qaeda leaders.[108] Again, this program focused upon the physical control of enemy leadership rather than its ideas, and no top-level operatives fell prey to this tactic. Thus, early PSYOP in OEF supported the kinetic campaign but yielded insignificant results.

Nonkinetic actions that provided indirect effects toward the ideological COG included a considerable humanitarian relief program. Leaflet drops outlining the nature and appearance of humanitarian rations occurred simultaneously with combat operations, and subsequent airdrops targeted areas identified by military planners as susceptible to food shortages. Over 2.5 million rations were eventually delivered, with no reports of starvation received during hostilities.[109] Furthermore, the humanitarian program expanded rapidly following major combat operations to incorporate air-land delivery, as well as terrestrial transport in support of UN operations.

Although statistics illustrating the psychological impact of these operations do not exist for Afghanistan, polling data in response to humanitarian operations in a permissive environment reflect positive psychological effects. Following the 2005 Pakistani earthquake disaster, "the percentage of Pakistanis with favorable views of the US [doubled] from 23 percent to 46 percent from May 2005 to November 2005."[110] A similar phenomenon occurred in Indonesia after the tsunami disaster in 2004. Furthermore, these gains in US image are short-lived and must be edified by programs of more permanence to maintain the effect. Such actions did not transpire in Afghanistan until US policy makers recognized the growing import of the psychological component of OEF in the summer of 2002.

The most direct contribution to the ideological COG by nonkinetic assets surrounded coalition support for the Afghani elections. Both the October 2004 presidential election and the September 2005 National Assembly and Provincial Council elections received attention from the coalition's Election Information Group (EIG) to ensure Afghani understanding of the democratic process. First, the EIG developed message themes emphasizing the elections' security, legitimacy, transparency, and inevitability. Taking these ideas, the Joint Psychological Operations Task Force developed "posters, booklets, handouts, and radio and television broadcasts to encourage the Afghans to participate in the elections." Furthermore, one infantry brigade even constructed a "traveling road-show to help instruct the Afghans on the electoral process."[111] Ground- and air-based kinetic assets provided visible assurances of voter protection. Although these coalition nonkinetic efforts benefited from kinetic support to ensure some degree of security, they nonetheless provided tangible contributions to the ideological COG by enabling Afghanistan to experience its first democratically elected government in over 30 years. This synergy between kinetic and nonkinetic programs focused upon the relevant COG ought to represent the standard of US military operations, not the exception.

The evolution of the PRT program embodies the most significant US expression of the growing need to address OEF's psychological component with nonkinetic activities as the supported effort. Originally very small and limited in scope, PRTs evolved to perform as "integrated U.S. teams of military civil affairs units, political officers from the U.S. Embassy, and USAID [US Agency for International Development]-funded assistance teams, which were spread around the countryside."[112] These teams were intended to "monitor [local] conditions and progress [and to] facilitate, coordinate, and deliver humanitarian, reconstruction, and developmental aid."[113] When properly employed, PRTs extend US mission-commander and indigenous-government influence "into the far-reaching corners of the nation" while "portray[ing] . . . concern for the welfare and prosperity of the affected population, as opposed to combat soldiers who are often viewed as occupiers."[114] As emphasis in OEF shifted to support the PRT concept, the PRT mission coalesced around the provision of security and reconstruction, as well as the extension of national government influence. Each of the prongs of the PRT trident proved positive in puncturing the impenetrable ideological COG that had been mystifying US strategists.

The PRT program finds its origin in General Franks's decision to erect a Coalition Joint Civil-Military Operations Task Force (CJCMOTF) at the onset of OEF. History shows such an organization is not designated automatically, and the very recognition of the requirement for a CJCMOTF indicates US admission to the importance of the psychological element in OEF. However, the fact that the combatant commander provided no guidance as to the organization's specific tasks or perceived mission reveals the initial low priority of effort devoted to the CJCMOTF. This phenomenon may be attributable to the lack of overarching message guidance from senior civilian echelons. Directed only to follow the lead of the UN, the CJCMOTF established small teams of coalition humanitarian liaison cells (CHLC) that were tasked "to assess humanitarian needs, conduct small reconstruction projects, and establish a relationship with the United Nations Assistance Mission in Afghanistan [UNAMA]."[115] Comprised of military civil-affairs soldiers, CHLCs populated extant UN humanitarian centers and SOF-controlled areas to facilitate immediate humanitarian needs with no mandate to consider post-conflict reconstruction.[116]

Following the local backlash associated with the increase in US conventional operations in the summer of 2002, lower-level military officials recognized the requirement to expand the scope and organizational composition of the CHLCs. Renamed joint regional teams (JRT), the new entities contained not only civil affairs expertise but also incorporated civil-military operations centers and security forces to enable activity in the ever-growing nonpermissive environment.[117] JRT composition remained military-centric, with integration with civilian agencies largely an afterthought. In November 2002, President Karzai asked that JRTs be designated PRTs with the astute political observation that "warlords rule regions; governors rule provinces."[118] At this stage, PRT configuration resembled that illustrated in figure 3, often with only a single, junior civilian among its membership. Throughout this nascent period, the PRT concept suffered from struggles for relevancy, flexible funding, and organizational harmony. The ambiguous mission enveloping the PRT idea contributed to its tentative quest for significance. Claims that PRTs should "monitor, . . . assist coordinating bodies, . . . facilitate cooperation," and the like imparted the impression that the teams were "all things to all people" but not actually accomplishing anything vital to the political or military missions.[119] Furthermore, the most tangible PRT contributions were confined to Kabul, further limiting the credibility and relevancy of the program. PRT efforts remained subservient to kinetic operations and were considered a "civil affairs thing" distinct from OEF's main effort.

Archaic funding restrictions designed for a peacetime bureaucracy limited the scope and integration of early CHLC/JRT/PRT projects. Initial CJCMOTF projects fell under Overseas Humanitarian Disaster Civil Aid (OHDCA) funding, requiring justification under its extensive guidelines. The OHDCA rules "demanded intense involvement of much of the CJCMOTF and CHLC staffs," which limited their ability to focus upon

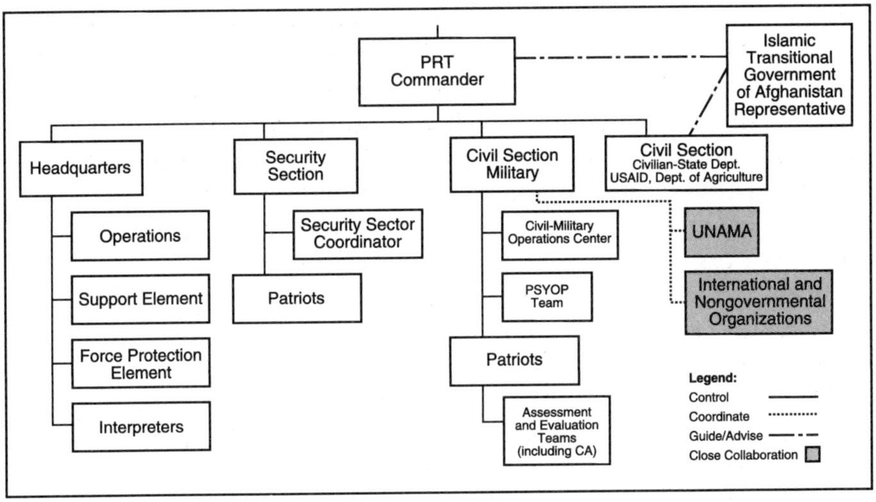

Figure 3. Early PRT structure in Afghanistan. (Reprinted from Col John D. Drolet, "Provincial Reconstruction Teams: Afghanistan vs. Iraq—Should We Have a Standard Model?" US Army War College Strategy Research Project [Carlisle, PA: US Army War College, 15 March 2006], 6.)

other mission areas.[120] Once the many layers of coordination met with approval, projects were offered to local contractors for bidding, and the local economy ultimately supplied the labor and materials for the effort. Funding became such a problem for the CJCMOTF that its vehicle fleet came to resemble "a post-apocalyptic Mad Max movie," while the initial OHDCA outlay of $2 million took five months to arrive; all this transpired with no relief from operational funds.[121]

Furthermore, the historical precedent for the use of OHDCA funds required emphasis upon "quick impact, high profile projects to jumpstart humanitarian civic action," forcing the CJCMOTF to target smaller-scale projects in a haphazard, disintegrated fashion.[122] Major infrastructure ventures did not comply with OHDCA strictures, compelling the CJCMOTF to pursue inflexibly school, clinic, and well repairs at the expense of road, electricity, or security (prison) endeavors. This practice contributed to preliminary PRT efforts comprising "many small disparate projects . . . not designed [as] a part of larger systems" of national reconstruction.[123] Stemming from this, civil-military relations exhibited signs of strain in the formative stages of the PRT concept.

The overwhelming military presence within early PRTs, coupled with the absence of joint civil-military training prior to deployment, contributed to organizational tension. Structurally, PRTs contained dozens of military personnel and were commanded by a lieutenant colonel. Civilian agencies often offered a single individual to liaise with the PRT, typically a junior-level diplomat with other competing duties. "Civilians complained that the military personnel . . . were reluctant to support them and treated them as outsiders. Military personnel were discouraged that civilians showed up with no resources, little authority . . . and sometimes little understanding of their role."[124] The experience often represented the first time a civilian relief worker had ever functioned within a military operation. PRT leadership was also contentious, with some civilians expressing frustration with the seemingly narrow focus of military PRT commanders, while some military commanders considered civilian contributions toward mission direction and emphasis lacking. The British-run PRT seemed to solve each of these issues by training together prior to deployment and adopting a "first among equals" philosophy regarding the military command structure.[125] Despite the initial growing pains, PRT civil-military relations improved with time and additional representation.

By December 2002, higher-level military leadership began to realize the worth of the PRT concept and its potential to leverage the ideological COG within OEF. Gen Richard Myers, chairman of the Joint Chiefs of Staff, announced a shift in emphasis away from combat operations and toward reconstruction efforts on 6 December 2002. Suggesting that the time to "flip" military priorities may be at hand, General Myers acknowledged that "the reconstruction piece in Afghanistan" should comprise the supported effort rather than the "hunting down of al Qaeda and Taliban fighters."[126] The significance of this admission of misguided emphasis cannot be overstated, as it illustrates the Hegelian struggle within US military strategy over the appropriate kinetic and nonkinetic weights of effort.

This strategic breakthrough stimulated a massive influx of resources and responsibility into the PRT program. Funding no longer presented a significant obstacle as the PRT budget expanded to $12 million and gained access to both DOS Economic Support Funds and the Commander's Emergency Response Program (CERP). These funding sources were not encumbered by strict employment guidelines, and PRT projects expanded immediately and flexibly into larger areas of increased national importance.[127] PRTs burgeoned to 60–100 soldiers during 2003, with civilian presence also increasing to include representatives from the DOS, USAID, and the Department of Agriculture, with an Afghani interior minister added in 2004. Both military and civilian tour lengths expanded to at least a year to facilitate organizational cohesiveness and continuity. PRT presence expanded throughout the nation as well, with four full PRTs operating by July 2003, eight by November, 18 by late 2004, and 22 in 2006, to incorporate "every significant populated area."[128]

Lt Gen David Barno removed the stigma associated with the nonkinetic mission of the PRTs by grouping them under the control of regional brigade commanders. He also mimicked the civil operations and revolutionary development support model from Vietnam by dispatching small PRT detachments to live in outlying villages rather than in the large, fortified coalition bases.[129] The international legitimacy of the PRT mission also increased, with five populated by NATO personnel by the fall of 2004, and four by International Security Assistance Force (ISAF) members by the summer of 2005. This rapid expansion received positive feedback from indigenous populations that welcomed the additional security and employment at more dispersed, localized environments.[130]

Although this accelerated evolution of the PRT brought with it clearer objectives, the organization's growing scope jeopardized its effectiveness. By January 2003, PRTs enjoyed the explicit mandate to extend the authority of the central government beyond Kabul, to contribute to the reconstruction process, and to enable civilian relief efforts through the provision of security and the establishment of relationships with local leaders.[131] However, PRT activities proliferated beyond these core tasks to include disarmament, police training, and de-mining, earning the comment from a senior military official that PRTs were becoming "a Christmas tree that everyone wants to hang their ornaments on."[132] General Barno reversed this trend by solidifying PRT taskings to the essentials of security, reconstruction, and the extension of governance. The flexible structure and varying locations of the PRTs allowed for variability in the priority each mission received, but their overall core competencies stabilized upon these three areas.

In the area of security, PRT performance is mixed. These organizations tend to experience success in mitigating outbreaks of local violence when a sound relationship exists between PRT representatives and local leaders. By assessing local conditions and capitalizing upon trust-based relationships with "local leaders of influence" who are typically current or erstwhile warlords, effective PRTs often defuse explosive confrontations before they devolve into violence.[133] PRT commanders often employ detailed organizational diagrams outlining tribal relationships in their area to leverage the appropriate figures during a particular disagreement.[134] While these interactions do well to avoid hostility, the ability of PRTs to quell violence, should mediation fail, is limited. The security forces personnel within PRTs are not combat units, and instances of PRTs unable to prevent continued violence between rival tribes do exist. Furthermore, some PRTs emphasize the reconstruction mission over their security role and exert little effort in forestalling factional in-fighting.[135] Thus, in the role of security provision, PRT performance receives varied reviews.

PRTs excel in focusing reconstructive efforts. By tailoring resource expenditure to the particular needs of an affected populace, these reconstruction efforts target precisely their popular opinion and therefore influence indirectly the ideological COG of OEF. Continual presence enables PRTs to become finely attuned to the vagaries of their locale, engendering intelligent assistance administered with a deft touch.[136] Furthermore, the growing integration of civilian elements into the PRTs, including central government officials, enhances unity of effort across the political, military, and nongovernmental spectrum. Measurement of the ideological gains associated with reconstructive efforts remains a conundrum. "The number of smiling Afghani children" has been cited as one indicator of PRT programs' positive impacts, but more stringent measures beg discovery.[137] Public opinion polls have been attempted but with little tangible success. Thus, PRT reconstruction projects provide positive contributions to the ideological COG, but quantifiable measurement methods should be pursued to focus further the effects of these efforts.

Of the current slate of PRT functions, the extension of central government legitimacy represents the most direct effect upon the ideological COG. By attributing PRT improvements to the Karzai administration, coalition nonkinetic campaigns contribute to long-term regional stability. "The importance of Karzai's election in this milieu cannot be underestimated," and his authority provides the foundation for an ideological alternative to the dictatorial Taliban.[138] By subtly lessening the local influence of regional chieftains through relationship building and visible reconstruction, PRTs offer a "firm basis to extend Afghani government influence."[139] This phenomenon represents a critical link between the nonkinetic PRT program and the ideological COG of the GWOT. Only by influencing local populations in such a nonbrute force method will the adverse ideological effects of the kinetic portion of OEF find absolution.

Thus, US military leadership came to recognize grudgingly the indicators necessitating an operational shift from the kinetic to the nonkinetic. Although the contributions of OEF's nonkinetic campaign toward the ideological COG are "clearly having a positive impact," they remain difficult to quantify.[140] Furthermore, reconstruction in Afghanistan continues to progress at a snail's pace; this may be attributable to a variety of confounding factors. Afghanistan's war-torn history represents one reason for the difficulties of current reconstruction efforts, although present capability in certain areas remains worse than before the start of OEF. The change of military focus from Afghanistan to Iraq in the summer of 2002 may have slowed recognition of the need for a nonkinetic emphasis. Not only did OIF pull resources and attention from Afghanistan, it also forced NATO and ISAF to assume a greater leadership role. Regardless of the myriad contributors to the delayed nonkinetic programs in OEF, the PRT concept harbors hope for the attainment of US political objectives in Afghanistan.

Operation Iraqi Freedom

Preliminary indications in OIF suggest US military strategy continues to struggle in balancing war's physical and moral elements.

The Kinetic Model

OIF commenced on 19 March 2003 with decapitation air strikes targeting Hussein-regime leadership. Ground operations followed 18 hours later with an emphasis on speed that enabled the president to declare the cessation of major combat operations by 1 May 2003. Driven to deny the nexus of "radicalism and technology" in the form of WMD programs in Saddam Hussein's Iraq, the United States adopted a kinetic-centric military strategy targeting the Iraqi COGs of leadership, personnel (Republican Guard), and weaponry (WMDs).[141] The United States, tired of Hussein's ambiguity regarding his weapons programs and military action, exuded a compellent stance uninterested in compromise or negotiation. The more extensive infrastructure and target sets of Iraq enabled the prosecution of a more destructive campaign than OEF in terms of resource expenditure, but increased levels of civilian leadership interaction with military planning resulted in an expedient decision not to apply overwhelming force. The choice of economy over dominance proved detrimental to long-term stability as the sectarian-based Iraqi insurgency began to bloom. As in Afghanistan, the United States elected to step up its kinetic efforts in the face of sectarian violence despite evidence that its actions contributed to diminished popular support of both the US and Maliki governments. An ever-increasing level of violence, however, is the most telling indicator regarding the questionable nature of America's kinetic-centric operations. As in OEF, short-term physical gains were impressive, but eventually US policy makers came to admit that their kinetic focus needed amendment.

The Nonkinetic Model

As in OEF, the administration acknowledged the presence of an ideological element, but nonkinetic efforts remained a secondary concern. US public diplomacy programs experienced little progress, with structural changes occurring too late and with slight effect. Weak US attempts to espouse a message continued to focus on its internal audience, and OIF suffered from strategic ambiguity even more so than OEF. Initiated under the guise of the GWOT to eliminate stores of extremist-accessible WMDs, OIF transformed into a multifaceted counterinsurgency desperate to instill a stable democracy. "Ambiguous goals never promote strategic coherence," and as such, the Bush administration continued to grasp for consistency in its public diplomacy message.[142] Bush established the Muslim World Outreach Policy Coordinating Committee in July 2004 to refine further American messages transmitted to the Arab community. Furthermore, the DOS added an additional layer to its bureaucracy to assist in the allocation of its public-diplomacy resources and planning efforts.[143] Nonetheless, a 2006 Government Accountability Office report concluded that the US government "continues to lack . . . an interagency communications strategy to guide government-wide public diplomacy activities," and the nation's inability to handle the scandal of Abu Ghraib exposed this deficiency.[144] Furthermore, an April 2007 assessment of American public diplomacy "improvements" concludes that "the score is not in our favor" and that "with a sustained and focused strategy, . . . with patience and perseverance, we should begin to see the fruits of our labor in the years to come."[145] Although optimistic, such an observation after almost six years of conflict under the GWOT, along with claims that "global opinion of U.S. foreign policy has sharply deteriorated in the past two years," brings into question the validity and seriousness of State Department efforts to bolster the US public diplomacy program since 9/11.[146]

Militarily, unclassified informational efforts continue to focus upon America's domestic audience and the support of kinetic operations. The Pentagon-sponsored Office of Media Outreach funds travel for selected US media personnel "to report the good news on OIF . . . to get the news straight from our troops . . . , including the positive developments and successes they are achieving."[147] These programs do little to target the COG of ideology within the GWOT and, instead, attempt to bolster the longevity of US domestic support for the war. PSYOP in the field attempted to influence Iraqi forces to capitulate, refrain from employing WMDs, and avoid damaging the infrastructure of the nation's oil industry. Military PSYOP succeeded in these campaigns; however, only the oil-preservation component of these nonkinetic actions arguably "set the conditions for the events that would follow the collapse of the regime."[148] Furthermore, for oil revenues to contribute to the ideological COG, the funds must be linked to tangible infrastructure improvements that garner Iraqi support for the Maliki government. So far, evidence of this relationship does not exist.

Other military OIF efforts to persuade Muslim audiences continue to be shrouded in surreptitiousness. A September 2004 DOD contract proposal sought to address declining support for US actions in Iraq. Iraqex, a two-individual firm, won the $5.4 million deal and immediately changed its name to the Lincoln Group. The business history of the firm's founders include "a string of short-lived businesses such as Express Action, an Internet-based shipping company that raised $14 million in startup financing during the dot-com boom but disappeared within two years; [and] Motion Power, an attempt to invent a shoe that would generate electrical power."[149] In Iraq, the Lincoln Group "dabbled in real estate, published a short-lived online business publication called the *Iraq Business Journal*, and tried its hand at exporting scrap metal, manufacturing construction materials, and providing logistics for US forces before finally striking gold with the Pentagon PR [public relations] contract."

The initial contract expanded to $57.6 million, and on 30 November 2005, the Lincoln Group's purpose came to light. The *Los Angeles Times* reported that the US military was "secretly paying Iraqi newspapers to publish stories written by American troops in an effort to burnish the image of the U.S. mission in Iraq." The articles were "written by US military IO troops, translated into Arabic, and placed in Baghdad newspapers" by the Lincoln Group to mask any connection with the military.[150] Although the stories were found to be "basically factual," they tended to present only the US side of an issue. It is unfortunate that the absence of focused national-level public diplomacy, coupled with American adherence to World War II propaganda concerns expressed in the Smith-Mundt Act, forces the DOD to rely upon suspect individuals to perform a critical informational function. The US populace should recognize the worth of such activities and allow the United States to erect organizations to administer international persuasion efforts more effectively.

The PRT concept has taken root in Iraq, however, and this amalgamation of kinetic and nonkinetic implements may halt eventually the hemorrhaging of US popular opinion and bring credibility to the Maliki government. Modeled after the OEF PRT experiment, the Iraqi version materialized as a provincial support team (PST) in April 2005, an amazing *three years* following the concept's initial use in Afghanistan. Only six to eight individuals comprised the initial PSTs, with each team led by a DOS official. Incorporating "a military deputy, a representative from the [USAID], Iraqi Reconstruction Management Office field activity, a unit combat engineer, a unit civil affairs officer, and a representative from the Army . . . responsible for security in that province," initial PSTs pursued a purely advisory role in Iraqi reconstruction efforts and languished under lukewarm institutional support.[151]

Gen George W. Casey Jr., commander of Multi-National Forces-Iraq, infused the program with $80 million of CERP funds to jump-start PST effectiveness. CERP dollars had already subsidized short-term improvement projects that served to "win hearts and minds twice over—once by repairing infrastructure, and again by employing local citizens who are otherwise ready recruits for the insurgents."[152] These funds coupled with the PST organizational link to local provinces enabled projects to become not only more attributable to the Iraqi government but also more integrated into a national recovery plan. Early success of this construct engendered a massive expansion of the PST concept during the fall of 2005 that witnessed the burgeoning of the teams to approximately 100 personnel and their redesignation as PRTs.[153] The PRT role continues to expand, providing positive contributions to reconstruction, governmental legitimacy, and integrated policy. Each of these outcomes supports indirectly the ideological COG of the GWOT.

Effects

America's kinetic military strategy in Iraq, along with the seemingly successful removal of al-Qaeda's safe-haven COG in

Afghanistan, generated two particularly alarming effects. First, the forced dispersal of al-Qaeda leadership from Afghanistan served to "franchise" Islamism in other regions.[154] This decentralization of activity throughout the al-Qaeda network presents significant difficulties for a strategy focused upon kinetic search-and-destroy tactics. Second, the al-Qaeda movement now exhibits decentralization of control as well. This phenomenon enables al-Qaeda to inspire disparate adherents to its ideological cause with no positive direction or funding. Internal US government assessments describe this effect as the "metastasizing of jihadist terrorism," and its prevalence can be directly attributable to the United States' lack of emphasis upon countering the extremist message.[155] These characteristics of the al-Qaeda movement coupled with incessant US efforts to corral and kill its leaders allow the continual replenishment of its manpower pool.

Clumsy, destructive attacks pursuant to the leadership COG in Iraq often result in heavy civilian casualties that contribute to the insurgency's popularity. One particular leadership strike involved six JDAMs striking a single villa. Although the individual targeted was not at home, collateral damage claimed a key tribal leader and ally of the CIA along with 21 members of his family. This leader represented an informal figurehead of an extended tribe comprising as many as two million Iraqis.[156] The long-term impact of such callous military activity amidst an ideological conflict remains to be discovered, although resentment stemming from this act undoubtedly exists. Furthermore, experts contend that a strategy deemphasizing economic and infrastructure improvements contributes to an aura of "misery and resentment" that accelerates al-Qaeda recruitment.[157] Without targeting the ideological inspiration and physical discontentment of potential Islamist recruits, US kinetic efforts are bound to flounder. This realization propelled Secretary Rumsfeld to inquire early into OIF whether the United States was "capturing, killing, or deterring and dissuading more terrorists every day than the madrassas and the radical clerics [were] recruiting, training and deploying against us."[158] Apparently, staffers convinced the secretary of an affirmative answer, providing fuel for the continuation of America's misguided kinetic strategy.

The destabilizing effects of America's kinetic-centric strategy in OIF are evident in figure 4 and table 3, which demonstrate the spiraling trend of violence in Iraq despite a surge in US kinetic-based efforts to provide population security. Recent data continues the ominous trend with "the number of American troops killed by homemade bombs in Iraq . . . nearly doubl[ing] . . . since the 'surge' of forces began . . . this spring [2007]."[159] Claims that "the invasion of Iraq ha[s], as a leaked National Intelligence Estimate noted, made the overall problem of jihadist terrorism worse than it was before 2003" garner greater veracity over time, and US popular opinion abroad fares no better.[160] "Recent polls show the image of the US declining throughout the world and that large majorities of Muslim populations believe the US seeks to undermine Islam as a religion."[161] It seems the military strategy undergirding OIF placed more emphasis on destroying enemy forces, eradicating WMDs, and forcing regime change than setting the stage for the long-term ideological war.

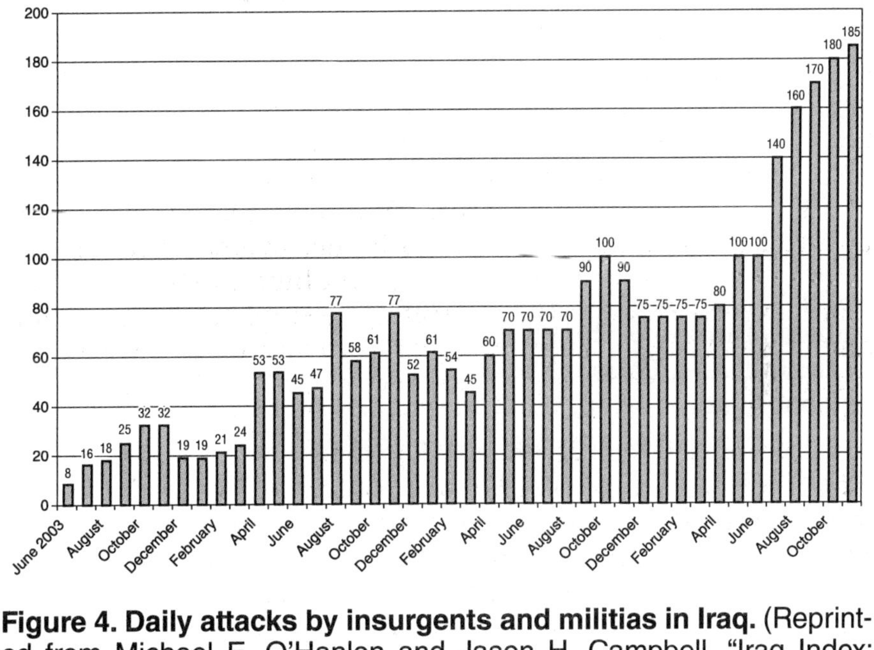

Figure 4. Daily attacks by insurgents and militias in Iraq. (Reprinted from Michael E. O'Hanlon and Jason H. Campbell, "Iraq Index: Tracking Variables of Reconstruction and Security in Post-Saddam Iraq" [Washington, DC: Brookings Institution, 7 May 2007, 28].)

Table 3. Iraqi security trends

Year	1 May 03–19 Mar 04	20 Mar 04–19 Mar 05	20 Mar 05–19 Mar 06	20 Mar 06–16 Mar 07
Number of Days	324	365	365	362
Civilians Killed per Day	20	31	41	73
Civilians Killed	6,332	11,312	14,910	26,540
– By Mortars (Number of Attacks)	30 (10)	303 (82)	286 (73)	1,374 (289)
– In Car/Suicide/Roadside Attacks (Number of Attacks)	742 (55)	1,782 (276)	3,402 (488)	5,797 (976)
– In Bombings Killing More Than 50 (Number of Attacks)	447 (5)	446 (6)	712 (9)	1,476 (17)

Adapted from "Year Four: Simply the Worst," Press Release 15, Iraq Body Count, 18 March 2007, http://www.iraqbodycount.org/press/pr15.php. These figures do not include the 7,400 civilian deaths recorded during the six-week invasion phase in March–April 2003.

In particular, the United States failed to understand that the foundation of governmental credibility that leads to ideological legitimacy "lies not in the destruction of the old system, but in the creation of the new one."[162] Once the moral component of the conflict gained greater preeminence in the minds of US policy makers, positive programs such as the PRT concept gained momentum. However, the late start and lack of planning for implementation of the PRTs forced them to tackle a much more daunting task. Thus, as in OEF, America's kinetic-centric strategy chose short-term, temporary gains over long-term ideological achievements, serving to fuel unintentionally the extremists' cause.

Notes

1. This study references the most recent versions of the other relevant documents as they reflect changes specifically in response to 9/11.
2. *National Security Strategy*, 2002, 1.
3. Ibid.
4. DOD, *National Defense Strategy*, 6.

5. *National Military Strategy*, iv, 1.

6. *National Military Strategic Plan for the War on Terrorism*, 12–13.

7. DOD, *Quadrennial Defense Review Report*, 6 February 2006, 21–22.

8. Ibid., 22.

9. *National Military Strategic Plan for the War on Terrorism*, 19. Admittedly, one may debate the logic leading to the GWOT's political objectives of "defeat violent extremism," and "create a global environment inhospitable to violent extremists and all who support them" in support of the *NSS* objectives of freedom, prosperity, and liberty.

10. *National Security Strategy*, 2002, 5.

11. *National Military Strategic Plan for the War on Terrorism*, 13.

12. Hoffman, *Inside Terrorism*, *19*.

13. Biddle, *American Grand Strategy after 9/11*, 7–8.

14. Bush, "Address to a Joint Session of Congress."

15. Lewis, *Crisis of Islam*, 137.

16. Pipes, "Bin Laden Is a Fundamentalist."

17. Although mainstream Muslims may object to the term "Islamism" as representative of an extremist strain of their religion, its use has become increasingly accepted.

18. National Commission on Terrorist Attacks, *9/11 Commission Report*, 50.

19. See this study's appendix for a more detailed presentation of the origins of Islamist thought and its influence upon al-Qaeda's worldview and political agenda.

20. Despite its recent 2006 publication date, the *NMSP-WOT* merely elaborates upon the ideas of earlier strategic documents and expresses characteristics of violent extremist organizations that transcend time. Therefore, its use to frame the COGs of the early stages of the GWOT is not unfounded.

21. *National Military Strategic Plan*, 14–19, 24–27.

22. DOD, *Quadrennial Defense Review Report*, 6 February 2006, 21.

23. *National Security Strategy*, 2002, 5–6, 31.

24. Bush, "Presidential Address to the Nation."

25. The White House, *Global War on Terrorism*.

26. Senate, *Statement of General Tommy Franks*, 4.

27. Aldape, "US Military Campaign in Afghanistan."

28. Ibid.

29. Wills, "Airpower, Afghanistan," 23–24.

30. Williams, "Post Conflict Reconstruction," 170.

31. O'Hanlon, "Flawed Masterpiece," 51.

32. Biddle, *Military Power*, 200.

33. O'Hanlon, "Flawed Masterpiece," 51–52.

34. Aldape, "US Military Campaign in Afghanistan."

35. O'Hanlon, "Flawed Masterpiece," 53.

36. Garamone, "Afghan Campaign Entering New Phase"; and RAND National Defense Research Institute, *Operation Enduring Freedom*, 1.

37. Aldape, "US Military Campaign in Afghanistan."

38. RAND National Defense Research Institute, *Operation Enduring Freedom*, 1.
39. Ibid.; and Williams, "Post Conflict Reconstruction," 171.
40. Aldape, "US Military Campaign in Afghanistan."
41. Weisman, "Raids Intensify."
42. Pandya, *Security, Reconstruction*, 7.
43. Williams, "Post Conflict Reconstruction," 167.
44. O'Hanlon, "Flawed Masterpiece," 51, 59.
45. Pandya, *Security, Reconstruction*, 3.
46. Williams, "Post Conflict Reconstruction," 169.
47. Urrutia-Varhall, *Public Diplomacy*, 22.
48. O'Hanlon, "Flawed Masterpiece," 54.
49. Conetta, *Operation Enduring Freedom.*
50. RAND National Defense Research Institute, *Operation Enduring Freedom*, 2.
51. Aldape, "US Military Campaign in Afghanistan."
52. Pandya, *Security, Reconstruction*, 2.
53. Wafa, "U.S. Says Raids Killed Taliban."
54. White, "69 Afghans' Families Get a U.S. Apology."
55. Hendrickson, "Toward Universal Empire," 8–9.
56. *Afghanistan Country Overview 2006.*
57. Williams, "Post Conflict Reconstruction," 193.
58. Rheinheimer, "Afghanistan Stage III."
59. Maloney, "Afghanistan Four Years On," 31.
60. Pandya, *Security, Reconstruction*, 1.
61. Williams, "Post Conflict Reconstruction," 172.
62. Maloney, "Afghanistan Four Years On," 31.
63. Ford, "Europe Cringes at Bush 'Crusade.' "
64. Biddle, *American Grand Strategy after 9/11*, 8.
65. Muravchik, "Hearts, Minds," 25.
66. Ibid., 26.
67. *Improving the U.S. Public Diplomacy Campaign*, 2.
68. Kilroy, "Public Diplomacy," 131.
69. *Information Operations Newsletter*, 29.
70. Johnson and Dale, "How to Reinvigorate US Public Diplomacy," 7.
71. Ibid.
72. House, *Statement of Lisa A. Curtis.*
73. Muravchik, "Hearts, Minds," 28.
74. Johnson and Dale, "How to Reinvigorate US Public Diplomacy."
75. Government Accountability Office (GAO), *US Public Diplomacy: Interagency Coordination Efforts*, 10.
76. GAO, *US Public Diplomacy: State Department Efforts*, ii.
77. GAO, *US Public Diplomacy: State Department and Broadcasting*, 11.
78. *Improving the U.S. Public Diplomacy Campaign*, 1.
79. GAO, *US Public Diplomacy: State Department Efforts*, 5.
80. Muravchik, "Hearts, Minds," 28.
81. GAO, *US Public Diplomacy: State Department Efforts*, 5.

82. *Foreign Affairs Reform and Restructuring Act.*
83. Ungar, "Pitch Imperfect," 7.
84. *Smith-Mundt Act of 1948.*
85. Johnson and Dale, "How to Reinvigorate US Public Diplomacy."
86. Leigh Armistead, ed., *Information Operations*, 5.
87. Johnson and Dale, "How to Reinvigorate US Public Diplomacy." Radio Sawa means "Radio Together" in Arabic.
88. Senate, *Statement of Kenneth Y. Tomlinson.* Radio Farda translates to "Radio Tomorrow" in Persian.
89. Ibid.; and GAO, *US Public Diplomacy: State Department and Broadcasting*, 8.
90. Ungar, "Pitch Imperfect," 10. Al-Hurra means "The Free One" in Arabic.
91. GAO, *US Public Diplomacy: State Department and Broadcasting*, 9.
92. Ungar, "Pitch Imperfect," 11.
93. Lieven and Hulsman, *Ethical Realism*, 50.
94. GAO, *US Public Diplomacy: State Department Efforts*, 6.
95. House, *Statement of Lisa A. Curtis.*
96. GAO, *US Public Diplomacy: Interagency Coordination Efforts*, 16; and Johnson and Dale, "How to Reinvigorate US Public Diplomacy."
97. Muravchik, "Hearts, Minds," 25.
98. Kilroy, "Public Diplomacy," 131.
99. Cox, "Information Operations?" 59.
100. Urrutia-Varhall, *Public Diplomacy*, 19–21.
101. Peterson, "Public Diplomacy," 75.
102. Urrutia-Varhall, *Public Diplomacy*, 32.
103. *Improving the U.S. Public Diplomacy Campaign*, 1.
104. House, *Statement of Lisa A. Curtis.*
105. Alexander, *Advanced Weapons*, 184–89.
106. Urrutia-Varhall, *Public Diplomacy*, 19; and *Information Operations Newsletter*, 1.
107. Lynch, "Al-Qaeda's Media Strategies."
108. Cox, "Information Operations?" 57.
109. Williams, "Post Conflict Reconstruction," 167.
110. House, *Statement of Lisa A. Curtis.*
111. Cox, "Information Operations?" 60–62.
112. Pandya, *Security, Reconstruction*, 5.
113. Wardle, "Search for Stability," 4.
114. Drolet, "Provincial Reconstruction Teams," 4.
115. Ibid.
116. Williams, "Post Conflict Reconstruction," 190.
117. Drolet, "Provincial Reconstruction Teams," 5.
118. McNerney, "Stabilization and Reconstruction?" 36.
119. Ibid.
120. Williams, "Post Conflict Reconstruction," 191–92.
121. McNerney, "Stabilization and Reconstruction?" 36.
122. Williams, "Post Conflict Reconstruction," 191.

123. Drolet, "Provincial Reconstruction Teams," 13.
124. McNerney, "Stabilization and Reconstruction?" 37.
125. Wardle, "Search for Stability," 11.
126. Teeple, "US Military Shift Afghan Operation."
127. McNerney, "Stabilization and Reconstruction?" 37.
128. Maloney, "Afghanistan Four Years On," 22, 25.
129. McNerney, "Stabilization and Reconstruction?" 38.
130. Williams, "Post Conflict Reconstruction," 193.
131. Drolet, "Provincial Reconstruction Teams," 6.
132. McNerney, "Stabilization and Reconstruction?" 38.
133. Wardle, "Search for Stability," 7.
134. McNerney, "Stabilization and Reconstruction?" 40.
135. Pandya, *Security, Reconstruction*, 5.
136. Wardle, "Search for Stability," 14.
137. McNerney, "Stabilization and Reconstruction?" 39.
138. Maloney, "Afghanistan Four Years On," 24.
139. Ibid., 25.
140. McNerney, "Stabilization and Reconstruction?" 43.
141. *National Security Strategy*, 2002, 13.
142. Biddle, *American Grand Strategy after 9/11*, 28.
143. GAO, *US Public Diplomacy: Interagency Coordination Efforts*, 10.
144. GAO, *US Public Diplomacy: State Department*, ii; and Schultz and Dew, *Insurgents, Terrorists, and Militias*, 244.
145. House, *Statement of Lisa A. Curtis*.
146. Sullivan, "Views on US Drop Sharply."
147. *Information Operations Newsletter*, 29.
148. Cox, "Information Operations?" 66.
149. *Information Operations Newsletter*, 30.
150. Ibid., 31.
151. Drolet, "Provincial Reconstruction Teams," 7–8.
152. Nagl, *Learning to Eat Soup with a Knife*, xiii.
153. Drolet, "Provincial Reconstruction Teams," 9–10.
154. *National Military Strategic Plan for the War on Terrorism*, 13.
155. Gellman and Linzer, "Afghanistan, Iraq: Two Wars Collide," 421.
156. Schultz and Dew, *Insurgents, Terrorists, and Militias*, 234–35.
157. Biddle, *American Grand Strategy after 9/11*, 24.
158. Healy, "We're at War."
159. Lubold, "US Losses in Iraq."
160. Fukuyama, *America at the Crossroads*, x.
161. House, *Statement of Lisa A. Curtis*.
162. Kagan, "War and Aftermath," 10.

Chapter 6

Conclusion

The enemy can usually reintroduce violence, and therefore one must be prepared for such an eventuality.

—David Lonsdale

The analysis of any problem with as many confounding variables as America's global war on terror must attempt to sort through the miasma of data to discern some semblance of understanding. This study assaults this formidable task by erecting a framework of military effectiveness based upon Clausewitzian principles and then viewing US military strategy in Operation Enduring Freedom from both a kinetic and nonkinetic perspective. The simple fact that US military efforts have not attained the stated political objectives of the GWOT within OEF implies flawed logic behind the US military's penchant for physical, kinetic solutions to its ideological, political problems. Even more telling, however, is evidence of a Hegelian struggle within the US military—once it became aware of the ineffectiveness of its kinetic strategy.

The resistance to strategic change that US policy makers exhibited, despite growing evidence of the importance of OEF's psychological element, illustrates the powerful US military preference for kinetic operations. Furthermore, the military's incremental shift to a nonkinetic emphasis in OEF displayed painful slowness as an ill-equipped organization adapted to create unforeseen capabilities and relationships. The language of national-level strategy documents demonstrated increased recognition of the importance of the GWOT's ideological COG and acknowledged ideas as the sole strategic COG in a conflict that "can make new terrorists faster than [we] can capture or kill them."[1] This struggle over operational emphasis within the US military finds succinct expression in the latest draft of the US Marine Corps' *Small Wars* manual that suggests "if our political objectives can only be accomplished after a successful stability phase, then the stability phase is, de facto, the decisive phase."[2] This challenge to the traditional perception of kinetic, combat operations constituting the decisive phase reveals the internal struggle within the US military, spurred by the experiences of OEF.

The United States neglected the nonkinetic portions of its military strategy despite glaring indications of the primacy of the GWOT's ideological basis. Not only did the United States elevate the war's moral component through its characterization of the conflict as one against the subjective term "terrorism," it also discounted the profound ideological stimuli prevalent within al-Qaeda, forgetting that "terrorism's best asset . . . is the fire in the bellies of its young men [and women]."[3] Furthermore, US policy makers came to express a strategy of favoring kinetic actions targeting leadership and safe-haven COGs in the short-term while acknowledging that long-term success required nonkinetic operations leveraging the ideological COG.[4] However, efforts to alter military emphasis displayed an ad hoc nature, and focused informational strategies to energize the moderate Muslim community remained conspicuously absent.

This tendency of military emphasis to shift over time is not accounted for by Colonel Darley's interpretation of Clausewitz's continuum of violence. Although Darley suggests conflicts can be placed statically and in their entirety along the spectrum, the experiences of OEF imply that operational emphasis upon a war's physical and moral elements may, and perhaps should, shift as real-time indicators demand. Kinetic operations in OEF to remove the Taliban regime undoubtedly contributed to the "creation of a global environment inhospitable to violent extremists and all who support them"; therefore, an early strategy targeting the physical elements of the Taliban seems appropriate, despite the overarching importance of the ideological COG. Furthermore, once conditions suggested kinetic operations were in fact more detrimental than helpful in the pursuit of political objectives, nonkinetic methods rightly received greater weight.

Another shortfall of the continuum of violence stems from the assertion that as wartime violence approaches the abstract maximum, the physical dimension of conflict assumes greater importance. However, increased incidences of particularly violent resistance within OEF enhanced US perception of the primacy of the ideological element. Furthermore, Abu Mussab al Zarqawi's cries for excessive violence as elections neared during OIF illustrate this tendency.[5] This suggests that as war's moral element increases, so does the degree of violence. In Darley's defense, his claim is based upon the scale, scope, and duration of a conflict, not necessarily the character of any particular violent act. Nonetheless, assuming that a conflict favoring the moral realm will exude lower degrees of violence may prove perilous. Figure 5, below, may offer a more accurate depiction of the relationship between physical and psychological factors in war, and Darley's conception is reproduced as well to simplify comparison (fig. 6). Capitalizing upon economic theory, figure 6 illustrates an optimization curve representing the ideal collection of military capabilities. Furthermore, the optimized force comprised of both kinetic and nonkinetic capabilities must also possess the ability to move along the curve depending upon the character of a particular conflict.

Finally, a common perception among American policy makers suggests that "the political costs of underreaction are always going to be higher than the costs of overreaction."[6] Therefore, despite of the primacy of a given war's psychological components, the United States must prepare for politically-feasible

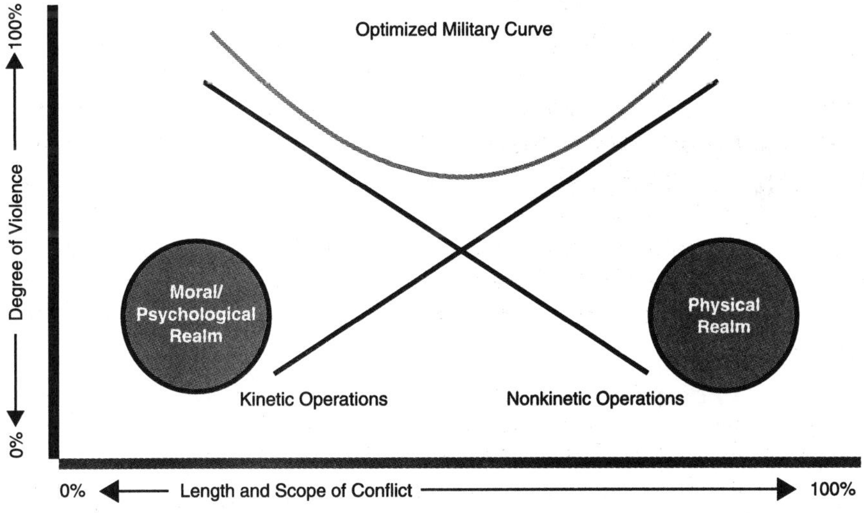

Figure 5. Alternative relationship of violence to warfare

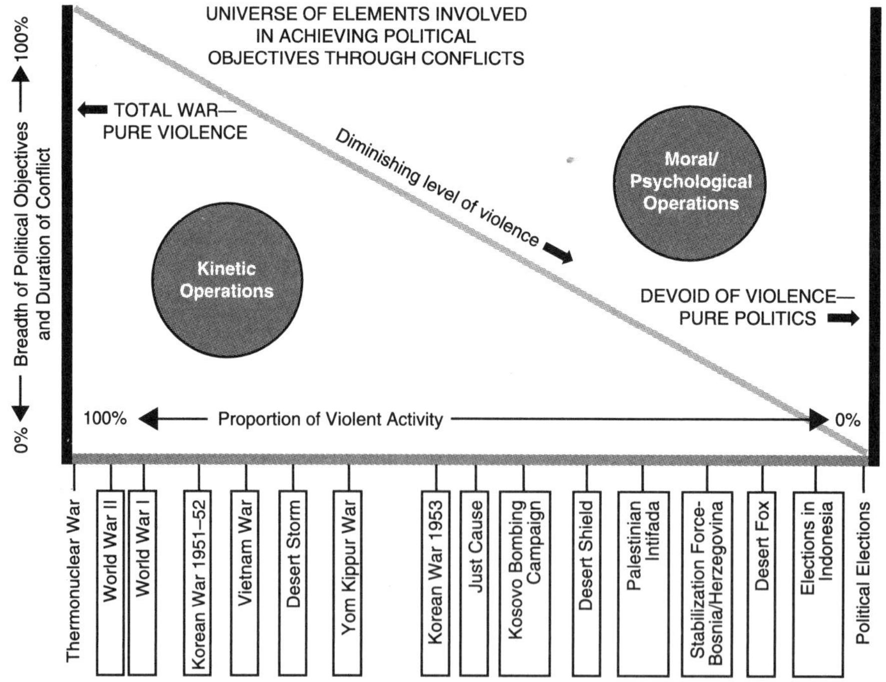

Figure 6. The continuum of violence. (Reprinted from Col William M. Darley, briefing, US Army Combined Arms Center, Fort Leavenworth, KS, subject: Information Operations Theory and Public Affairs, April 2004, slide 61.)

methods of combat that may dictate a kinetic, physical focus. Thus, regardless of a conflict's perceived character regarding its physical and moral components in advance of hostilities, a balanced force capable of favoring either dimension seems ideal.

This flexible, balanced force represents the synthesis of the Hegelian dialectic apparent in OEF. However, although Hegel's thesis and antithesis represent abstract opposites, the synthesis of the kinetic and nonkinetic models comprises a complementary, synergistic relationship. The need to exploit COGs in wartime with both kinetic, physical means as well as nonkinetic, psychological means is becoming generally acknowledged. However, institutional recognition of the need to shift flexibly to emphasize either complement of capabilities depending upon the changing character of a conflict is only now becoming manifest. Given that this deficiency garners greater attention only as OEF and OIF progress, the US military organization stands ill-equipped to fulfill this vision. It finds itself "rooted in an American approach to war that is singularly ill-fitted to the purposes of political reconstruction."[7] Furthermore, the US military's nonkinetic efforts will continue to lack strategic effectiveness without the existence of a unified message emanating from national leadership. Therefore, to evolve beyond the characteristics of the extant kinetic model, the US military must overcome its "cultural bias toward kinetic solutions" in war, while the US government must exude greater capacity to transmit unambiguous strategic communications.[8]

Recommendations

In the kinetic realm, the United States must posture itself organizationally and doctrinally to recognize nonkinetic activities as the supported effort in some circumstances. Military planning must incorporate the expectation that a shift in operational emphasis may not only be desirable but inevitable. Furthermore, plans and capabilities to enact such a shift at the appropriate time must exist at the inception of conflict. Although the exact time and manner of an operational change of priority must remain flexible to the exigencies of the situation, waiting to initiate the move until after initial indications of a failed strategy, as in OEF, is unacceptable. Military guidance, particularly the *National Military Strategy* and joint and service doctrine, needs to reflect the institutional desire to possess the capability to feature nonkinetic capabilities when required. The kinetic flavor of the current *NMS* leaves little room for nonkinetic effects.

Preliminary indications throughout the US military imply acceptance of this nonkinetic requirement. The 2006 *QDR* includes plans to expand PSYOP and CA units by 33 percent; identifies "capability gaps in each of the primary supporting capabilities of Public Affairs, Defense Support to Public Diplomacy, Military Diplomacy, and Information Operations"; and acknowledges military support to stabilization and reconstruction efforts "as a core mission."[9] Furthermore, General Casey, the current Army chief of staff, now emphasizes the need for America's ground arm to prepare for "persistent combat" that incorporates a larger proportion of nonkinetic capabilities. Similarly, the current Central Command commander, Adm William Fallon, rejected recently the term *long war* to refocus efforts upon reconstruction.[10]

The US military should also strive to maintain congruency with political guidance and resist the tendency to operate in isolation from politics. Despite the perceived failure of elevated civilian involvement in military affairs during OIF, military operations divorced from policy guidance represent a bankrupt basis for effective strategy. Reforms in the public diplomacy arena should facilitate the fusion of political vision with military means, granting greater congruency to both the kinetic and nonkinetic portions of US military strategy.

Regarding public diplomacy within OEF, the United States should soften the compellent tone of its message. "Attempting to sound resolute and strong . . . fails to recognize that many nations exist in politically-conflicted reality" and leaves little room for meaningful compromise leading to reconciliation.[11] This assertion does not profess to "negotiate with terrorists"; rather, it seeks to acknowledge that Muslim nations represent critical resources in the ideological struggle of the GWOT, and all applicable state venues should be explored. A January 2007 Gallup World Poll discovered that "Muslims and Americans generally agreed on the need to control extremism"; therefore, increased efforts to search for practical ways to engage with the Muslim world to build upon our shared values must be attempted.[12] As Joseph Nye contends, US political objectives can be achieved in the GWOT only "if moderate Muslims win [the] civil war between moderates and extremists within Islamic civilization. . . . We need to adopt policies that appeal to moderates, and to use public diplomacy more effectively to explain our common interests."[13]

National-level public diplomacy reform must occur to properly guide military informational efforts. These reforms must "strengthen the White House role in coordinating messages for international audiences and to provide a context for DOD wartime communications."[14] Although the Bush administration has already taken positive steps in this area, the sluggish responsiveness of the US bureaucracy needs to be assaulted with renewed vigor. Furthermore, America needs to awaken to the reality that, stemming from economic and media globalization, foreign and domestic propaganda grows progressively more difficult to distinguish. The DOD's 2003 IO road map acknowledges this fact, noting that "information intended for foreign audiences, including public diplomacy and PSYOP, increasingly is consumed by our domestic audience and vice-versa."[15] This fact should not alarm the majority of a US public already bombarded by politically-skewed news reports and rampant advertising. A serious review of the outdated Smith-Mundt Act of 1948 must occur, and should the nation decide change is warranted, an OSI-type organization represents an ideal construct to further focus national public diplomacy for military purposes.

The hodgepodge that IO has become should be separated. Broad interpretations of what exactly comprises IO must incorporate virtually every system and mission that involves the transmission of data. This comprehensive array of capabilities touches upon every mission of the US military, afflicting those responsible for overarching IO concepts of operation with sensory overload. Since the technical aspects of IO contribute to the full spectrum of military operations, the systems-based elements of IO should be disaggregated and integrated across the military specialties. Conversely, the psychological elements of IO, which are not reliant upon technological solutions, should be centralized and elevated in organizational stature. These moves will ensure the integration of those IO capabilities best suited to support kinetic operations, while at the same time freeing the informational piece of IO to focus upon the military's message and remain proactive in the war of ideas. Consolidation of the nonkinetic portions of IO under a joint force information component commander within the joint air operations center construct may prove ideal to achieving the desired flexibility between kinetic and nonkinetic operational emphases.

Finally, the positive experiences and tremendous potential of the PRT concept as a microcosm of the kinetic/nonkinetic synthesis mandates its preservation. The PRT's combination of security, reconstruction, and the extension of indigenous governance comprises a powerful collection of capabilities with which to extend influence. Furthermore, its conduciveness to scalability and its ability to adapt to contextual environments and coalition requirements enhance the PRT as an ideal capability for an uncertain future. The integrated nature of the PRT's personnel complement fits well with coalition and civilian aims to achieve unity of effort in the administration of humanitarian programs. The experience of OEF's provincial reconstruction teams must be codified within doctrine and tactics, techniques, and procedures manuals to maintain its efficacy in future conflicts.

OEF reveals also areas for PRT improvement. The security element of the PRT mission set may receive greater enforcement capability to enable reconstructive and governance functions to occur earlier within a conflict. Additional muscle may also allow the incorporation of US public diplomacy officials into the PRT membership to alleviate their continual "struggle to balance security with public access and outreach to local populations."[16] Furthermore, the required civil-military relationships of the PRT construct should be maintained in peacetime if feasible. Training, job exchanges, and simple two-way communication between organizations all support smoother operations once hostilities transpire, and increased civilian involvement may engender a larger frequency of civilian leadership of future PRTs—a sensible organizational construct given the PRT's nonkinetic emphasis. At the very least, civil-military training programs in preparation of deployment must be erected and maintained. Establishing measures of effectiveness comprise the most compelling area for PRT improvement. Although most recognize the positive impact of PRTs, increased institutional acceptance of their important role will be facilitated by tangible expressions of their effect.

Thus, the kinetic-centric US military strategy within OEF exhibits shortcomings in the attainment of political objectives. Fortunately for the coalition effort, the US military came to realize that long-term success required a shift in operational emphasis. Given that the United States itself was born of a revolutionary insurgency, it should strive to erect and preserve the capability to alternate its priority of military effort between kinetic and nonkinetic means. Finally, the establishment of a coherent, national-level strategic communications architecture must be mastered to provide the consistent message military informational efforts so desperately crave.

Notes

1. *National Military Strategic Plan for the War on Terrorism*, 21.
2. McNerney, "Stabilization and Reconstruction?" 33.
3. Palmer, *Terrorism, War, and the Press*, 10.
4. *National Strategy for Combating Terrorism*, 7.
5. Schultz and Dew, *Insurgents, Terrorists, and Militias*, 245–46.
6. Ignatieff, *Lesser Evil*, 58.
7. Hendrickson, "Toward Universal Empire," 9.
8. Murphy, "Information Operations," 2; and House, *Statement of Lisa A. Curtis*.
9. DOD, *Quadrennial Defense Review Report*, 6 February 2006, 45, 92, A-4.
10. Scarborough, "New Chief Wants Army"; and Lardner, "Message-Minded Admiral."
11. Kilroy, "Public Diplomacy," 135.
12. House, *Statement of Lisa A. Curtis*.
13. Nye, *Soft Power*, 131.
14. Johnson and Dale, "How to Reinvigorate US Public Diplomacy."
15. *Information Operations Newsletter*, 29.
16. GAO, *US Public Diplomacy*, 9.

Appendix

Al-Qaeda Background

Derived from the prevailing Muslim view of Islam as both a faith and a code of conduct, a standard for religion as well as politics, Islamism finds its roots within the terminal phase of Western colonialism. Arguably first embodied by the Muslim Brotherhood, an organized expression of Muslim dissatisfaction with the Western-backed secular monarchy of 1928 Egypt, Islamism bases its core doctrine upon Brotherhood member and executed political dissident Seyyid Qutb's interpretations of the Qur'an.[1] Capitalizing upon the Islamic term *jahiliyya* that represents ignorance of the Prophet Mohammed's revelations, Qutb describes Western, secular society as *jahili*, fraught with materialism, perversion, and disobedience to Allah. In opposition to *jahiliyya* stands Islam, the true path, complete submission to God; humanity must choose between the two. Given the base attraction of mankind to the physical comforts of *jahiliyya*, all Muslims must not only struggle personally (internal jihad) to stay the course of Islam but also combat outwardly (external jihad) the spread of adverse *jahili* influences. Finally, Westerners as well as Muslims who do not take up this charge as a sacred duty are considered nonbelievers and worthy of jihad aggression.[2]

A large school of thought chalks up Qutb's radical prose to blatant misrepresentations of the Qur'an. However, just as current interpretations of the Holy Bible remain subject to the distortions of modernity, the Qur'an is "a vast, vague book, filled with poetry and contradictions. . . . In it you can find condemnations of war and incitements to struggle, beautiful expressions of tolerance and stern strictures against unbelievers."[3] Therefore, one cannot confront Islamism with proof of Qur'anic misreading; rather, debate to discredit Islamism must focus upon its distortion of the core tenets of Islam.

The Islamist vision of a devout Muslim generates significant appeal among and within Islamic society for several cultural and ideological reasons. As a means to generate popular appeal, Islamists cite centuries past of Islamic regional domination—a

Golden Age when their ideology and culture spread throughout the Arabian Peninsula, the Middle East, North Africa, and even Europe.[4] Experienced at a time recently removed from Mohammed himself, this era of prosperity and influence seems to coincide with the Muslim culture's greatest degree of congruency with the Prophet's teachings. Therefore, the Islamists assert, a proactive jihad to restore a fundamentalist transnational community should culminate in a similar period of glory based on Allah's favor.

This message found receptive ears during the mid-twentieth century as many secular regimes within the Middle East failed to deliver upon promises of economic, technological, and even cultural advancement. As Muslim societies wallowed in poverty under governments with Western ties, Islamic fundamentalist groups offered charity and humanitarian aid to find audience for their beliefs.[5] Furthermore, Western influence in the region represented by globalization and the proliferation of modern materialism came to typify a cultural assault upon the Islamic faithful, with the United States materializing as its most blatant manifestation.

Osama bin Laden adopted this worldview during his religious studies at the University of Saudi Arabia and while listening to the "fiery sermons of Abdullah Azzam, a Palestinian and disciple of Qutb."[6] Inspired to confront perceived Western aggression and fulfill his Islamist calling, bin Laden volunteered to join and ultimately help finance the Afghani mujahideen and their jihad against the Soviet Union in the 1980s. The decade saw not only the expansion and solidification of an Islamist organization but also witnessed its triumph over a superior, technological foe in April 1988. Formalizing the victorious collection of jihadists under the moniker *al-Qaeda*, meaning "base," bin Laden worked to sustain the Islamist cause through recruitment and the selection of a new adversary.

Bin Laden and al-Qaeda issued a series of fatwas during the 1990s outlining their grievances against the United States. These serve as approximations of their motivations for conflict. Briefly, a *fatwa* is nothing more than a legal ruling within Islamic Shari'a law, typically issued by legitimate jurists and based upon historical precedent and consensus opinion.[7] The fact that al-Qaeda presumes to harbor the authority to offer legal judgments despite the dearth of Islamic scholars within its community represents a significant indicator of the movement's extremist and heretical nature.[8] Regardless, al-Qaeda's initial fatwa against the United States in 1992 demanded American expulsion from Somalia. Subsequent sporadic bombings and al-Qaeda training of Somali forces served to fulfill the fatwa with the United States' ignominious departure in 1994.

Emboldened by this success and coupled with minimal US responses to other terrorist attacks, including the World Trade Center in 1993, bin Laden delivered a public fatwa in 1996 of greater significance. Calling for jihad against Western occupation of Islamic lands, bin Laden specifically cites US military presence in Saudi Arabia as an egregious example of encroaching *jahili* influence.[9] US/Saudi relations began to sour for a multitude of reasons during this time, and by August 2003, America abandoned its Saudi positions.

Bin Laden's most recent and most expansive fatwa, signed in 1998 by a coalition of international Islamists including the Jihad Group in Egypt and its leader, Ayman al-Zawahiri, explicitly declares war upon the United States based upon American activities in the region that "are a clear declaration of war on God, his messenger, and Muslims." Again citing Qur'anic authority, bin Laden and the World Islamic Front state that the killing of "Americans and their allies—civilians and military" is "an individual duty for every Muslim who can do it in any country in which it is possible to do it."[10] Bin Laden clarified his extremist position months later during an interview on ABC television, claiming his stance of indiscrimination regarding military and civilian targets found precedent in the US atomic bombing of Japan.[11]

Thus, not only does al-Qaeda embody Islamism and its violent, extremist ideology, the movement also considers Western secularism, globalization, and modernization as the aggressors in a decades-long, possibly centuries-long, confrontation. As such, the events of 9/11 did not trigger the genesis of a new conflict as the declaration of a GWOT might imply. Rather, the attacks of 9/11 represent "reaction and revenge, even retribution" from al-Qaeda's perspective and simply the response to an unspoken invitation founded in a series of nominal American reactions to previous acts of terrorism.[12]

Notes

1. National Commission on Terrorist Attacks, *9/11 Commission Report*, 466.
2. Qutb, *Milestones*, 117–20, 139–40; and National Commission on Terrorist Attacks, *9/11 Commission Report*, 50–51.
3. Zakaria, "Why Do They Hate Us?" 117.
4. National Commission on Terrorist Attacks, *9/11 Commission Report*, 50.
5. Milton-Edwards, *Islamic Fundamentalism*, 118.
6. National Commission on Terrorist Attacks, *9/11 Commission Report*, 55.
7. Lewis, *Crisis of Islam*, 140.
8. National Commission on Terrorist Attacks, *9/11 Commission Report*, 47.
9. Ibid., 59–63.
10. World Islamic Front, "Statement: Jihad against Jews and Crusaders," 203.
11. "Hunting Bin Laden," PBS *Frontline* interview.
12. Zakaria, "Why Do They Hate Us?" 117; and Leibstone, "Strategic Communications," 6.

Abbreviations

BBG	Broadcasting Board of Governors
CA	civil affairs
CAIR	Council on American-Islamic Relations
CERP	Commander's Emergency Response Program
CFR	Council on Foreign Relations
CHLC	Coalition Humanitarian Liaison Cells
CI	counterintelligence
CJCMOTF	Coalition Joint Civil-Military Operations Task Force
CMOC	civil-military operations centers
CNA	computer network attack
CNO	computer network operations
COG	center of gravity
DOD	Department of Defense
DOS	Department of State
EIG	Election Information Group
EW	electronic warfare
GAO	Government Accountability Office
GPS	Global Positioning System
GWOT	global war on terror
IED	improvised explosive devices
IO	information operations
IPI	international public information
ISAF	International Security Assistance Force
JDAM	Joint Direct Attack Munition
JRT	joint regional team
MILDEC	military deception
NATO	North Atlantic Treaty Organization
NCW	network-centric warfare
NDS	*National Defense Strategy*
NGO	nongovernmental organization
NMS	*National Military Strategy*
NMSP-WOT	*National Military Strategic Plan for the War on Terrorism*
NSS	*National Security Strategy*

ODS	Operation Desert Storm
OEF	Operation Enduring Freedom
OHDCA	Overseas Humanitarian Disaster Civil Aid
OIF	Operation Iraqi Freedom
OPSEC	operations security
OSI	Office of Strategic Influence
PA	public affairs
PDD	Presidential Decision Directive
PRT	provincial reconstruction team
PST	provincial support team
PSYOP	psychological operations
QDR	*Quadrennial Defense Review*
RMA	revolution in military affairs
SOF	special operations forces
SSTR	stability, security, transition, and reconstruction
TRG	The Rendon Group
UN	United Nations
USAID	US Agency for International Development
USIA	US Information Agency
VOA	Voice of America
WMD	weapon of mass destruction

Bibliography

Academic Papers

Bass, Carla D. *Building Castles on Sand? Ignoring the Riptide of Information Operations*. Air War College, Maxwell Paper no. 15. Maxwell AFB, AL: Air University Press, August 1998.

Cox, Maj Joseph L. "Information Operations in Operation Enduring Freedom and Iraqi Freedom—What Went Wrong?" School of Advanced Military Studies Monograph. Fort Leavenworth, KS: US Army Command and General Staff College, 25 May 2006.

Drolet, Col John D. "Provincial Reconstruction Teams: Afghanistan vs. Iraq—Should We Have a Standard Model?" US Army War College Strategy Research Project. Carlisle Barracks, PA: US Army War College, 15 March 2006.

Furr, Ronald W. *Information Warfare: The Face of Future War, or a Rubric for Force Enhancement?* School of Advanced Airpower Studies. Maxwell AFB, AL: Air University Press, June 1995.

Mets, David R. *The Long Search for a Surgical Strike: Precision Munitions and the Revolution in Military Affairs*. CADRE Paper no. 12. Maxwell AFB, AL: Air University Press, October 2001.

Moorman, CDR John R. "The Future Role of Information Operations in Operational Art." Naval War College Paper. Newport, RI: Naval War College, 13 May 2002.

Shevchuk, CDR Benjamin A. "Military Implications of Moral Objectives in *The National Security Strategy of the United States of America*." US Army War College Strategy Research Project. Carlisle Barracks, PA: US Army War College, 18 March 2005.

Theisen, Eric E. *Ground-Aided Precision Strike: Heavy Bomber Activity in Operation Enduring Freedom*. Air War College Maxwell Paper no. 31. Maxwell AFB, AL: Air University Press, July 2003.

Wardle, Col Russell N. "The Search for Stability: Provincial Reconstruction Teams in Afghanistan." US Army War College Strategy Research Project. Carlisle Barracks, PA: US Army War College, 3 May 2004.

Watts, Barry D. *Clausewitzian Friction and Future War*. McNair Paper no. 52. Washington, DC: Institute for National Strategic Studies, National Defense University, October 1996.

Whitehead, Yulin G. *Information as a Weapon: Reality vs. Promises*. School of Advanced Airpower Studies. Maxwell AFB, AL: Air University Press, June 1997.

Williams, Col Garland H. "Post Conflict Reconstruction: On the Critical Path to Long-Term Peace." US Army War College Strategy Research Project. Carlisle Barracks, PA: US Army War College, 7 April 2003.

Wills, Lt Col Craig D. *Airpower, Afghanistan, and the Future of Warfare: An Alternative View*. CADRE Paper no. 25. Maxwell AFB, AL: Air University Press, November 2006.

Articles

Aylwin-Foster, Brig Nigel. "Changing the Army for Counterinsurgency Operations." *Military Review* 85, no. 6 (November–December 2005): 2–15.

Cebrowski, VADM Arthur K. "Network-Centric Warfare: Its Origin and Future." *Proceedings* 124, no. 1 (January 1998): 28–35.

Darley, Col William M. "Clausewitz's Theory of War and Information Operations." *Joint Forces Quarterly*, no. 40 (1st Quarter 2006): 73–79.

Ford, Peter. "Europe Cringes at Bush 'Crusade' against Terrorists." *Christian Science Monitor*, 19 September 2001. http://www.csmonitor.com/2001/0919/p12s2-woeu.html.

Garamone, Jim. "Afghan Campaign Entering New Phase." *Armed Forces Press Service*, 21 December 2001. http://www.defenselink.mil/news/newsarticle.aspx?id=44327.

Grant, Rebecca. "In Search of Lawful Targets." *Air Force Magazine*, February 2003, 38–44.

Gray, Colin S. "How Has War Changed since the End of the Cold War?" *Parameters* 35, no. 1 (Spring 2005): 14–26.

Healy, Gene. "We're at War Not with a State but an Armed Ideology." Cato Institute, 12 September 2004. http://www.cato.org/pub_display.php?pub_id=2828.

Hendrickson, David C. "Toward Universal Empire: The Dangerous Quest for Absolute Security." *World Policy Journal* 19 (Fall 2002): 1–10.

Hoffman, Frank. " 'New' Ideas That Look Old." *Armed Forces Journal*, July 2006, 29–31.

Jablonsky, David. "US Military Doctrine and the Revolution in Military Affairs." *Parameters* 24, no. 3 (Autumn 1994): 18–36.

Johnson, Stephen, and Helle Dale. "How to Reinvigorate U.S. Public Diplomacy." *The Heritage Foundation*, Backgrounder no. 1645, 23 April 2003. http://www.heritage.org/Research/PublicDiplomacy/bg1645.cfm.

Kagan, Frederick W. "War and Aftermath." *Policy Review*, no. 120 (August–September 2003): 3–27.

Kilroy, Richard J., Jr. "Public Diplomacy: Government, Universities, and the War on Terrorism." *Journal of Public Affairs* 8 (2005): 123–44.

Krepinevich, Andrew. "Keeping Pace with the Military Technical Revolution." *Issues in Science and Technology*, 22 June 1994, 23–29.

Lardner, Richard. "Message-Minded Admiral Ditches 'Long War' Phrase." *Tampa Tribune*, 19 April 2007.

Leibstone, Marvin. "Strategic Communications—Global Security and Being Understood." *Military Technology* 29, no. 9 (September 2005): 6.

Lubold, Gordon. "US Losses in Iraq Spike from IED Attacks." *Christian Science Monitor*, 23 May 2007.

Lynch, Mark. "Al-Qaeda's Media Strategies." *The National Interest Online*, 1 March 2006. http://www.nationalinterest.org/PrinterFriendly.aspx?id=11524.

Maloney, Sean M. "Afghanistan Four Years On: An Assessment." *Parameters* 35, no. 3 (Autumn 2005): 21–32.

McNerney, Michael J. "Stabilization and Reconstruction in Afghanistan: Are PRTs a Model or a Muddle?" *Parameters* 35, no. 4 (Winter 2005–06): 32–46.

Munro, Neil. "How Private is Your Data." *Washington Technology*, 9 February 1995, 14–16.

Muravchik, Joshua. "Hearts, Minds, and the War against Terror." *Commentary* 113, no. 5 (May 2002): 25–30.

O'Hanlon, Michael E. "A Flawed Masterpiece." *Foreign Affairs* 81, no. 3 (May/June 2002): 47–63.

Owens, Mackubin Thomas. "Technology, the RMA, and Future War." *Strategic Review* 26, no. 2 (Spring 1998): 63–70.

Peterson, Peter G. "Public Diplomacy and the War on Terrorism." *Foreign Affairs* 81, no. 5 (September/October 2002): 74–94.

Pipes, Daniel. "Bin Laden Is a Fundamentalist." *National Review Online*, 22 October 2001. http://www.nationalreview.com/comment/comment-pipes102201.shtml.

Ricks, Thomas E. "Leave Politics to Us, Warner Tells General." *Washington Post*, 24 January 2007. http://www.washingtonpost.com/wpdyn/content/article/2007/01/23/AR2007012301306.html.

Scales, Maj Gen Robert H., USA, retired. "Clausewitz and World War IV." *Armed Forces Journal*, July 2006, 16–24.

Scarborough, Rowan. "New Chief Wants Army to Prepare For 'Persistent Combat.'" *Washington Examiner*, 20 April 2007.

Schleifer, Ron. "Democracies, Limited War and Psychological Operations." *Review of International Affairs* 2, no. 3 (Spring 2003): 41–53.

Stokes, Bruce. "Public Diplomacy: America Is Job No. 1." *National Journal* 37, no. 19 (7 May 2005): 1–3.

Sullivan, Kevin. "Views on US Drop Sharply in Worldwide Opinion Poll." *Washington Post*, 23 January 2007.

Thomas, Timothy L. "Deterring Information Warfare: A New Strategic Challenge." *Parameters* 26, no. 4 (Winter 1996–97): 81–91.

Tolson, Jay. "The War of Ideas." *US News and World Report* 135, no. 8 (2003): 32.

Ungar, Sanford J. "Pitch Imperfect." *Foreign Affairs* 84, no. 3 (May/June 2005): 7–13.

Wafa, Abdul Waheed. "U.S. Says Raids Killed Taliban; Afghans Say Civilians Died." *New York Times*, 1 May 2007.

Weisman, Jonathan. "Raids Intensify; Red Cross Hit in Error." *USA Today*, 17 October 2001, final edition.

White, Josh. "69 Afghans' Families Get a U.S. Apology: Marines Killed 19 Civilians in March." *Washington Post*, 9 May 2007.

Books

Alexander, John B. *Advanced Weapons, Strategies, and Concepts for the Post-9/11 World*. New York: St. Martin's Press, 2003.

Allard, C. Kenneth. "Information Warfare: The Burden of History and the Risk of Hubris." In *In Athena's Camp: Preparing for Conflict in the Information Age*, edited by John Arquilla and David Ronfeldt. Santa Monica, CA: RAND Corporation, 1997.

Armistead, Leigh, ed. *Information Operations: Warfare and the Hard Reality of Soft Power*. Washington, DC: Brassey's, Inc., 2004.

Arquilla, John, and David Ronfeldt. "Information, Power, and Grand Strategy: In Athena's Camp." In *The Information Revolution and National Security Dimensions and Directions*, edited by Stuart J. D. Schwartzstein. Washington, DC: Center for Strategic and International Studies, 1996.

Bailey, Thomas A. *A Diplomatic History of the American People*. 10th ed. Englewood Cliffs, NJ: Prentice-Hall, Inc., 1980.

Biddle, Stephen. *Military Power: Explaining Victory and Defeat in Modern Battle*. Princeton, NJ: Princeton University Press, 2004.

Brown, Michael L. "Information Warfare and the Revolution in Military Affairs." In *Seminar on Command, Control, Communications, and Intelligence*, Guest Presentations, Spring 1995. Cambridge, MA: Program on Information Resources Policy, Harvard University, January 1996.

Clausewitz, Carl von. *On War*. Edited and translated by Michael Howard and Peter Paret. Princeton, NJ: Princeton University Press, 1976.

Cooper, Jeffrey R. "Another View of Information Warfare." In *The Information Revolution and National Security Dimensions and Directions*, edited by Stuart J. D. Schwartzstein. Washington, DC: Center for Strategic and International Studies, 1996.

Davis, Norman C. "An Information-Based Revolution in Military Affairs." In *In Athena's Camp: Preparing for Conflict in the Information Age*, edited by John Arquilla and David Ronfeldt. Santa Monica, CA: RAND Corporation, 1997.

Dolman, Everett Carl. *Pure Strategy: Power and Principle in the Space and Information Age.* New York: Frank Cass, 2005.

Franks, GEN Tommy. *American Soldier.* New York: HarperCollins Publishers, Inc., 2004.

Friedrich, Carl J., ed. *The Philosophy of Hegel.* New York: Random House, Inc., 1954.

Fukuyama, Francis. *America at the Crossroads: Democracy, Power, and the Neoconservative Legacy.* New Haven, CT: Yale University Press, 2006.

Gaddis, John Lewis. *Surprise, Security, and the American Experience.* Cambridge, MA: Harvard University Press, 2004.

Gellman, Barton, and Dafna Linzer. "Afghanistan, Iraq: Two Wars Collide." In *Understanding the War on Terror*, edited by James F. Hoge Jr. and Gideon Rose. New York: Council on Foreign Relations, Inc., 2005.

Girard, Jolyon P. *America and the World.* Westport, CT: Greenwood Press, 2001.

Gordon, Michael R., and Gen Bernard E. Trainor. *Cobra II: The Inside Story of the Invasion and Occupation of Iraq.* New York: Pantheon Books, 2006.

Gouré, Daniel. "The Impact of the Information Revolution on Strategy and Doctrine." In *The Information Revolution and National Security Dimensions and Directions*, edited by Stuart J. D. Schwartzstein. Washington, DC: Center for Strategic and International Studies, 1996.

Hall, Wayne Michael. *Stray Voltage: War in the Information Age.* Annapolis, MD: Naval Institute Press, 2003.

Hanson, Victor Davis. *Carnage and Culture: Landmark Battles in the Rise of Western Power.* New York: Anchor Books, 2001.

Hoffman, Bruce. *Inside Terrorism.* New York: Columbia University Press, 2006.

Hoge, James F., Jr., and Gideon Rose, eds. *Understanding the War on Terror.* New York: Council on Foreign Relations, Inc., 2005.

Huntington, Samuel P. *The Soldier and the State: The Theory and Politics of Civil-Military Relations.* Cambridge, MA: Harvard University Press, 1985.

Ignatieff, Michael. *The Lesser Evil: Political Ethics in an Age of Terror.* Princeton, NJ: Princeton University Press, 2004.

Joes, Anthony James, ed. *Saving Democracies: U.S. Intervention in Threatened Democratic States*. Westport, CT: Praeger Publishers, 1999.

Judis, John B. *The Folly of Empire: What George W. Bush Could Learn from Theodore Roosevelt and Woodrow Wilson*. New York: Oxford University Press, 2004.

Kagan, Frederick W. *Finding the Target: The Transformation of American Military Policy*. New York: Encounter Books, 2006.

Keegan, John. *A History of Warfare*. London: Pimlico, 1994.

———. *The Iraq War*. Toronto: Vintage Canada, 2005.

Kennan, George F. *Memoirs: 1925–1950*. Boston: Knopf, 1977.

Kitfield, James. *Prodigal Soldiers: How the Generation of Officers Born of Vietnam Revolutionized the American Style of War*. Washington, DC: Potomac Books, Inc., 1995.

Kurzweil, Ray. *The Age of Spiritual Machines: When Computers Exceed Human Intelligence*. New York: Penguin Books, 1999.

Lewis, Bernard. *The Crisis of Islam: Holy War and Unholy Terror*. New York: Modern Library, 2003.

Lieven, Anatol, and John Hulsman. *Ethical Realism: A Vision for America's Role in the World*. New York: Pantheon Books, 2006.

Lonsdale, David J. *The Nature of War in the Information Age: Clausewitzian Future*. New York: Frank Cass, 2004.

Manstein, Erich von. *Lost Victories*. Edited and translated by Anthony G. Powell. Novato, CA: Presidio Press, 1982.

Millis, Walter. *Arms and Men*. New Brunswick, NJ: Rutgers University Press, 1956.

Milton-Edwards, Beverly. *Islamic Fundamentalism since 1945*. New York: Routledge, 2005.

Nagl, John A. *Learning to Eat Soup with a Knife: Counterinsurgency Lessons from Malaya and Vietnam*. Chicago, IL: University of Chicago Press, 2005.

Nasr, Vali. *The Shia Revival: How Conflicts within Islam Will Shape the Future*. New York: W. W. Norton & Company, Inc., 2006.

Nye, Joseph S., Jr. *Soft Power: The Means to Success in World Politics*. New York: Public Affairs, 2004.

Odom, William E. *America's Military Revolution: Strategy and Structure after the Cold War.* Washington, DC: American University Press, 1993.

Palmer, Nancy, ed. *Terrorism, War, and the Press.* Cambridge, MA: Joan Shorenstein Center on the Press, Politics, and Public Policy, John F. Kennedy School of Government, Harvard University, 2003.

Peters, Rudolph. *Jihad in Classical and Modern Islam.* 2nd ed. Princeton, NJ: Markus Wiener Publishers, 2005.

Phillips, David L. *Losing Iraq: Inside the Postwar Reconstruction Fiasco.* Boulder, CO: Westview Press, 2005.

Quester, George H. *American Foreign Policy: The Lost Consensus.* New York: Praeger Publishers, 1982.

Qutb, Seyyid. *Milestones.* Damascus, Syria: Dar al-Ilm, n.d.

Record, Jeffrey. *Dark Victory: America's Second War Against Iraq.* Annapolis, MD: Naval Institute Press, 2004.

Ricks, Thomas E. *Fiasco: The American Military Adventure in Iraq.* New York: Penguin Press, 2006.

Ryan, David. *US Foreign Policy in World History.* New York: Routledge, 2000.

Schelling, Thomas C. *Arms and Influence.* New Haven, CT: Yale University Press, 1966.

Schultz, Richard H., Jr., and Andrea J. Dew. *Insurgents, Terrorists, and Militias: The Warriors of Contemporary Combat.* New York: Columbia University Press, 2006.

Schwartz, Stephen. *The Two Faces of Islam: The House of Sa'ud from Tradition to Terror.* New York: Doubleday, 2002.

Schwartz-Barcott, T. P. *War, Terror and Peace in the Qur'an and in Islam: Insights for Military and Government Leaders.* Carlisle Barracks, PA: Army War College Foundation Press, 2004.

Sifry, Micah L., and Christopher Cerf, eds. *The Iraq War Reader: History, Documents, Opinions.* New York: Touchstone, 2003.

Snow, Donald M., and Eugene Brown. *United States Foreign Policy: Politics beyond the Water's Edge.* 2nd ed. Boston: Bedford/St. Martin's, 2000.

Strassler, Robert B., ed. *The Landmark Thucydides: A Comprehensive Guide to the Peloponnesian War.* New York: Touchstone, 1996.

Toffler, Alvin, and Heidi Toffler. *War and Anti-War: Survival at the Dawn of the 21st Century*. Boston: Little, Brown & Co., 1993.

Van Creveld, Martin. *Command in War.* Cambridge, MA: Harvard University Press, 1985.

———. *The Transformation of War: The Most Radical Reinterpretation of Armed Conflict since Clausewitz.* New York: Free Press, 1991.

Warden, John A., III. *The Air Campaign.* Lincoln, NE: Excel Press, 2000.

Weigley, Russell F. *The American Way of War: A History of United States Military Strategy and Policy.* Bloomington, IN: Indiana University Press, 1973.

Woodward, Bob. *Bush at War.* New York: Simon & Schuster, 2002.

———. *State of Denial.* New York: Simon & Schuster, 2006.

World Islamic Front. "Statement: Jihad against Jews and Crusaders." In *The Iraq War Reader: History, Documents, Opinions,* edited by Micah L. Sifry and Christopher Cerf. New York: Touchstone, 2003.

Wright, Lawrence. *The Looming Tower: Al-Qaeda and the Road to 9/11.* New York: Alfred A. Knopf, 2006.

Zakaria, Fareed. "Why Do They Hate Us?" In *Understanding the War on Terror,* edited by James F. Hoge Jr. and Gideon Rose. New York: Council on Foreign Relations, Inc., 2005.

Briefings/Point Papers/Memos/Messages

Aldape, Sofia. "The US Military Campaign in Afghanistan: The Year in Review." Washington, DC: Center for Defense Information, 10 October 2002. http://www.cdi.org/terrorism/afghanistan-one-year-later-pr.cfm.

Boyd, John R. "A Discourse on Winning and Losing." Unpublished collection of briefings on competitive strategy, August 1987. http://www.d-n-i.net/second_level/boyd_military.htm.

———. "Patterns of Conflict." One of a series of unpublished briefings in "A Discourse on Winning and Losing," December 1986. http://www.d-n-i.net/boyd/patterns_ppt.pdf.

Buchan, Glenn. "Information War and the Air Force: Wave of the Future? Current Fad?" RAND Issue Paper, RAND Corporation, March 1996.

Darley, Col William M. Briefing. US Army Combined Arms Center, Fort Leavenworth, KS. Subject: Information Operations Theory and Public Affairs, April 2004.

Myers, Gen Richard C. Chairman, Joint Chiefs of Staff. "Policy on Public Affairs Relationship to Information Operations." Memorandum, 24 September 2004.

Murphy, Dennis. "Information Operations and Winning the Peace: Wielding the Information Element of Power in the Global War on Terrorism." Center for Strategic Leadership Issue Paper, US Army War College, December 2005.

O'Hanlon, Michael E., and Jason H. Campbell. "Iraq Index: Tracking Variables of Reconstruction and Security in Post-Saddam Iraq." Washington, DC: Brookings Institution, 7 May 2007.

Rheinheimer, Francis. "Afghanistan Stage III: NATO's Most Ambitious Operation?" Washington, DC: Center for Defense Information, 15 May 2006. http://www.cdi.org/program/document.cfm?DocumentID=3428&from_page=../index.cfm.

"Year Four: Simply the Worst." Press Release 15. *Iraq Body Count,* 18 March 2007. http://www.iraqbodycount.org/press/pr15.php.

Government Documents

Air Force Doctrine Document 2-5. *Information Operations*, 11 January 2005.

Department of Defense (DOD) Directive 3000.05. *Military Support for Stability, Security, Transition, and Reconstruction (SSTR) Operations*, 28 November 2005.

DOD. *The National Defense Strategy of the United States of America.* Washington, DC: Office of the Secretary of Defense, March 2005.

———. *Quadrennial Defense Review Report.* Washington, DC: Office of the Secretary of Defense, 30 September 2001.

———. *Quadrennial Defense Review Report*. Washington, DC: Office of the Secretary of Defense, 6 February 2006.

Field Manual 3-13. *Information Operations: Doctrine, Tactics, Techniques, and Procedures*, 28 November 2003.

House. *Foreign Affairs Reform and Restructuring Act*. 105th Cong., 2nd sess. HR 1757, 27 January 1998.

Information Operations Newsletter. G3 Plans, Information Operations Branch. Offutt AFB, NE: US Army Strategic Command, September 2006.

Information Operations Primer: Fundamentals of Information Operations. US Army War College document. Carlisle Barracks, PA: US Army War College, January 2006.

Joint Chiefs of Staff. *Joint Vision 2020*. Washington, DC: US Government Printing Office, 2000.

Joint Publication 3-13. *Information Operations*, 13 February 2006.

Marine Corps Warfighting Publication 3-40.4. *Marine Air-Ground Task Force Information Operations*, 9 July 2003.

National Military Strategic Plan for the War on Terrorism. Washington, DC: Chairman of the Joint Chiefs of Staff, 1 February 2006.

The National Military Strategy of the United States of America. Washington, DC: Chairman of the Joint Chiefs of Staff, 2004.

The National Security Strategy of the United States of America. Washington, DC: The White House, September 2002.

The National Security Strategy of the United States of America. Washington, DC: The White House, March 2006.

National Strategy for Combating Terrorism. Washington, DC: The White House, September 2006.

Naval Warfare Publication 3-13. *Navy Information Operations*, June 2003.

9/11 Five Years Later: Successes and Challenges. Washington, DC: National Security Council, September 2006.

Smith-Mundt Act of 1948. Public Law 80-402. 80th Cong., 2nd sess., 27 January 1948.

US Army War College. *Information Operations Primer: Fundamentals of Information Operations*. Carlisle Barracks, PA: Department of Military Strategy, Planning, and Operations, January 2006.

US Information Agency. Brochure. Washington, DC: USIA Printing Office, October 1998.
US Marine Corps. *Small Wars Manual.* Washington, DC: Government Printing Office, 1940.
US War Department. Field Manual 100-5. *Tentative Field Service Regulations, Operations.* Washington, DC: Government Printing Office, October 1939.
———. *Field Service Regulations, United States Army.* Washington, DC: Government Printing Office, 1924.

Interviews

"Hunting Bin Laden." PBS *Frontline* interview, May 1998. http://www.pbs.org/wgbh/pages/frontline/shows/binladen/who/interview.html.

Reports

Afghanistan Country Overview 2006. Washington, DC: The World Bank, September 2006. http://www.worldbank.org.af.
Allison, Graham T., and Paul X. Kelley. *Nonlethal Weapons and Capabilities: Report of an Independent Task Force Sponsored by the Council on Foreign Relations.* New York: Council on Foreign Relations, 2004.
Biddle, Stephen D. *American Grand Strategy after 9/11: An Assessment.* Carlisle Barracks, PA: Strategic Studies Institute, April 2005.
Coffman, Lt Col Alfred M., Jr. *Strategic Environmental Assessment for Modernization Planning.* Report of the Strategic Planning Division, Directorate of Plans. Washington, DC: Headquarters USAF, 6 June 1994.
Conetta, Carl. *Operation Enduring Freedom: Why a Higher Rate of Civilian Bombing Casualties.* Project on Defense Alternatives Briefing Report #11. Cambridge, MA: The Commonwealth Institute, 18 January 2002. http://www.comw.org/pda/0201oef.html#8.
Council on Foreign Relations. *Improving the U.S. Public Diplomacy Campaign in the War against Terrorism.* Task Force

Report No. 38. New York: Council on Foreign Relations Press, 6 November 2001.

Echevarria, Antulio J., II. *Toward an American Way of War.* Carlisle Barracks, PA: Strategic Studies Institute, March 2004.

Gough, Susan L. *The Evolution of Strategic Influence.* Carlisle Barracks, PA: Army War College, 7 April 2003.

Government Accountability Office (GAO). *U.S. Public Diplomacy: Interagency Coordination Efforts Hampered by the Lack of a National Communication Strategy.* GAO Report 05-323. Washington, DC: GAO, April 2005.

———. *U.S. Public Diplomacy: State Department and Broadcasting Board of Governors Expand Post-9/11 Efforts but Challenges Remain.* GAO Report 04-1061T. Washington, DC: GAO, 23 August 2004.

———. *U.S. Public Diplomacy: State Department Efforts Lack Certain Communication Elements and Face Persistent Challenges.* GAO Report 06-707T. Washington, DC: GAO, 3 May 2006.

Iraq Study Group. *The Iraq Study Group Report: The Way Forward—A New Approach.* New York: Vintage Books, 2006.

Jensen, Richard M. *Information War Power: Lessons from Air Power.* Cambridge, MA: Program on Information Resources Policy, September 1997.

Libicki, Martin C. *What Is Information Warfare?* Washington, DC: Center for Advanced Concepts and Technology, August 1995.

Mazarr, Michael J., Jeffrey Shaffer, and Benjamin Ederington. *The Military Technical Revolution: A Structural Framework.* Washington, DC: Center for Strategic and International Studies, March 1993.

Metz, Steven. *Armed Conflict in the 21st Century: The Information Revolution and Post-Modern Warfare.* Carlisle Barracks, PA: Strategic Studies Institute, April 2000.

National Commission on Terrorist Attacks upon the United States. *The 9/11 Commission Report: Final Report of the National Commission on Terrorist Attacks upon the United States* (Authorized Edition). New York: W. W. Norton & Company, Inc., 2004.

Pandya, Amit A. *Security, Reconstruction and Political Normalization in Afghanistan—Falling Short.* Report of the Center for American Progress. Washington, DC: Center for American Progress, January 2004.

Rabasa, Angel, Peter Chalk, Kim Cragin, Sara A. Daly, Heather S. Gregg, Theodore W. Karasik, Kevin A. O'Brien, and William Rosenau. *Beyond al-Qaeda: Part 1, The Global Jihadist Movement.* Santa Monica, CA: RAND Corporation, 2006.

———. *Beyond al-Qaeda: Part 2, The Outer Rings of the Terrorist Universe.* Santa Monica, CA: RAND Corporation, 2006.

RAND National Defense Research Institute. *Operation Enduring Freedom: An Assessment.* Research Brief. Santa Monica, CA: RAND Corporation, 2005.

Steele, Robert David. *Information Operations: Putting the "I" Back into DIME.* Carlisle Barracks, PA: Strategic Studies Institute, February 2006.

Urrutia-Varhall, Linda R. *Public Diplomacy—Capturing the Information Terrain on the Way to Victory.* Maxwell AFB, AL: Air University Press, April 2002.

The White House. *The Global War on Terrorism: The First 100 Days.* White House Counterterrorism Report. Washington, DC: The White House, 20 December 2001. http://www.state.gov/s/ct/rls/wh/6947.htm.

Speeches/Hearings

Bush, Pres. George W. "Address to a Joint Session of Congress and the American People." United States Capitol, Washington, DC, 20 September 2001. http://www.whitehouse.gov/news/releases/2001/09/20010920-8.html.

———. "Presidential Address to the Nation." The Treaty Room, The White House, Washington, DC, 7 October 2001. http://www.whitehouse.gov/news/releases/2001/09/20010920-8.html.

House. Committee on Foreign Affairs. *Statement of Lisa A. Curtis, Senior Research Fellow, The Heritage Foundation: Hearing of the House Committee on Foreign Affairs, Subcommittee on International Organizations, Human Rights, and*

Oversight. 110th Cong., 1st sess., 26 April 2007. http://foreignaffairs.house.gov/110/cur0426 07.htm.

House. International Relations Committee, Subcommittee on Oversight and Investigations. *Testimony of Andrew Kohut.* 109th Cong., 1st sess., 10 November 2005. http://www.au.af.mil/au/awc/awcgate/congress/koh111005.pdf

Senate. Armed Services Committee. *Statement of General Tommy Franks, Commander in Chief, US Central Command.* 107th Cong., 2nd sess., 7 February 2002. http://www.globalsecurity.org/military/library/con gress/2002_hr/Franks.pdf.

Senate. Committee on Foreign Relations. *Statement of Kenneth Y. Tomlinson, Chairman, Broadcasting Board of Governors: Hearing of the Senate Committee on Foreign Relations.* 108th Cong., 1st sess., 27 February 2003.

Teeple, J. "US Military Shift Afghan Operation towards Reconstruction Efforts." *Voice of America*, 6 December 2002.

US GOVERNMENT PRINTING OFFICE: 2008—725-002

After you have read this research report, please give us your frank opinion on the contents. All comments—large or small, complimentary or caustic—will be gratefully appreciated. Mail them to the director, AFRI, 155 N. Twining St., Maxwell AFB, AL 36112-6026.

Ideas in Arms **Torkelson**

The Relationship of Kinetic and Ideological Means in America's Global War on Terror

Thank you for your assistance.

Cut along dotted line